Memory Hound

Tales of an Immunologist

Franco Celada

Lamprey Books
New York
2019

Lamprey Books - 1st Edition, New York, September 2019

The Anti-DNA Story (Chapter Six) was originally published in the journal LUPUS in 2010.

Author: Franco Celada (Franco.Celada@nyumc.org)
Editor and Art Director: Valentina Celada (valentinacelada@gmail.com)

All illustrations by Franco Celada, except the "Lymphocyte Doodles" by Umberto Eco, p.169-170 and "The Moving Party" drawing by Guido Celada, p.357.

ISBN-13: 978-1-7338692-0-1

To Antonella, whose presence
is felt with love every day

It is not knowing, but the love of learning,
that characterizes the scientific man

Charles Sanders Peirce

From: "Lessons from the History of Science: The Scientific Attitude" (c.1896), in Collected Papers (1931), Vol. 1, 20.

Contents

Chapter Loglines

1 PORCINI. *An ironic parable on preconception, poison of science and reason.*

2 BORIS. *The beauty of pure discovery: an antibody is able to cure an enzyme crippled by mutation.*

3 NARRAGANSETT. *The "science" of sailing vintage vessels: a time-stopping ritual, reenacting historical milestones.*

4 BLOCKHEAD. *Ulysses' lesson: the essential nature of a man's name and the dire consequences of losing it.*

5 GILBERT. *Science seeks to understand, diplomacy to persuade. A scientist, torn between the two worlds, realizes they will never meet.*

6 THE ANTI-DNA STORY. *Serendipity concurs in a student's first discovery, which will shape the rest of his life and career.*

7 OCTOBER '56. *Time loss and twisted medicine are the insufferable facts of the military draft in the 1950's. Humor and smart aleck-ness are the only means of survival.*

8 PICCAPIETRA. *Sequestered by the Red Brigades, a serologist expects the worst, until he is told the reason why he was chosen.*

Preface

LIFE is defined by four dimensions: the three coordinates of SPACE, plus TIME. All living beings are precarious, on the verge of becoming outcompeted and extinct, which is the worst possible fate in a world where the evolutionary reward is survival. A cascade of necessary consequences follows: all beings must build their defenses and, for these to be competitive, they need to collect information on the dangers. Menacing invaders come from any direction, forcing organisms to monitor all four coordinates. Our monitoring agents are the SENSES, and a multitude of SENSORS down the scale of size and complexity to the simplest single-cell organism. Together they patrol space all around. Life's only time-scanner is MEMORY. Immunological Memory, mounted first in primitive fish, and subsequently in all vertebrates, is the product of the Specific Response and a tool that makes the Immune System "cognitive" and able to learn. It is concerned with internal fluids and internal environment. Its life-saving record ranks among the body's principal defenses and shares blueprints with the precursors of Mental Memory developing in the nervous system.

As a researcher in Biology, I have been playing with concepts of survival and evolution throughout my life. But it is my work on *Immunological Memory* that explains my attraction to *Memory without Adjectives* and the eagerness to anchor my recollections on paper.

Writing this book fueled my curiosity about the memories of two highly developed systems of our body: the *Nervous* and the *Lymphatic*.

Contemplating this topic in my office at the *Hospital for Joint Diseases*, was like rallying with an invisible tennis trainer machine, shooting hypotheses at me, as if they were tennis balls.

> Ball n.1 says: "*Both memories work for SAVING LIFE: Evolution's mission*".
>
> Ball n.2: "*Parallel forces favor memory's IMPRINTING: INFLAMMATION, in the lymphatic realm; EMOTION in the brain*".
>
> Ball n.3: "*The critical moment of both memories' imprinting is the plucking of the OBJECT (Virus or Event) from the FLUX of life and preserving it in an existence OUT OF TIME*".
>
> Ball n.4 hits the writer, and says: "*Do not mistrust the breaking of the FLUX. It is an old dream of humanity to stop time. After all, Goethe's Faust, is willing to sacrifice his soul to freeze a moment in time and instructs the Devil:*
>
> *"When I say to the Fleeting Moment:*
> *'Linger a while - thou art so fair!'*
> *Then bind me in thy bonds undying,*
> *and my final ruin I will bear!"* [1]

1. Fortunately, the experiment was never done, because Faust never found the Moment he was looking for. I guess it was fortunate also for Mephistopheles: what

The long "brewing" of this book included visits to places and conversations with several friends, who have become "characters" of my book, from my childhood years, to medical school, my days in the army and then work. I took trips to Paris, Milan, Velate, Genoa, Geneva, Stockholm, Rome, Los Angeles, Phoenix, Providence, Madison County and New York. A great excuse to rekindle old friendships, the meetings to compare recollections were always a pleasure and confirmed my enthusiasm: I found our memories to be largely concordant, with occasional different gaps suggesting an individually selective process of memory.

As a writer and immunologist, I can state that Memory, in all its forms, is serious stuff. A real child of Elysium, Memory shows us the deeds and misdeeds of our past, a great help in the decision-making process throughout our entire life.

Memory is also connected to our identity. In fact, as I was gathering and writing some of my childhood memories for this book, I realized that the identification with my "self" as a younger man was only partial. Thus, for the oldest memories, in Chapter Ten - *Stefania*, the remembered is addressed in the third person, with the nickname *Francolino*, as if he were a person other than the author.

The "tales" in the book describe facts and events that happened to and around me during the period between 1935 and 2018, in several countries, with my family always in tow. As you can see in the table of contents, the chapters are not presented in chronological order, but according to the moment each memory coagulated, came to mind and "demanded" to be written. I do not know whether my recollections are random or influenced by something in my

---

if he had not been able to fulfill his promise, especially after Joshua had stopped the sun for hours!

subconscious and, in this ignorance, I preferred to pay respect to whim, instead of imposing a rational order.

Throughout the book, real scientific experiments are intermingled with the facts of life, with some possibility of mutual influences and disturbances. The result is not naked Science but Life- in- and with- Science.

I feel fortunate to have become a scientist at a time when Science, and thus the Philosophy of Science, was evolving. The occasional descriptions of experiments delve into the universe of cells and molecules, but also into the realms of scientific reasoning and scientific attitude.

One of the milestones in this direction is attained when the Immune System comes to be granted the status of Cognitive System, an unconscious mind with its own "language" and decision-making tools, thanks to the fruitful collaboration with Umberto Eco.

Before I became a physician and then a scientist, my education has been classic and humanistic, with great emphasis on philosophy. A personal surprise was finding that translating from ancient Greek is a perfect training for scientific reasoning (see the reflections on Abduction in Chapter 9 - *Riding with Eli*), confirming that the Italian *Liceo Classico* (the Italian high school of Humanities) provides a better preparation to Science than *Liceo Scientifico* (science high school).

History buffs will notice there is a large section devoted to the years of World War II, experienced by children and adults alike, and set as the background of my forming years. The war, like a vaccine, had the effect of shaping young minds into freedom and justice lovers, certainly impermeable to any dictatorship and "sovereignties" that we see popping up all around today (Trump and his wall in the USA, the Brexit gang in the UK, Orbán in Hungary, Bolsonaro in Brazil, Salvini, chaser of refugees, in Italy).

Because these chapters, entitled *Sfollati*, ("Displaced") were so vivid in my memory and so "visual" on the page, I was encouraged by friends to write a screen adaptation entitled *Gli Sfollati del Sacro Monte* ("The Evacuees of the Holy Mountain"), which may become a script and perhaps a film in the future.

I want to reassure my old friends that the person they knew from school, or university has not changed, and even though the book touches on serious topics, they will find moments of humor, in the same spirit of the doodles we passed around under the benches during the long lectures of Hygiene at our Medical School in Milan.

Illustrations and doodles are a family legacy and have accompanied me throughout my life, in my lectures and also in the lab as a way of keeping humor always alive. Several of the drawings in the book are "historical" and can be recognized by the date shown, the remainder have been created for this book. It has been a pleasant surprise finding myself in good company when we discovered Eco's discarded doodles at a meeting in Lucca.

Finally, writing in English came spontaneously to me, having spent more than half of my life abroad with English being the international working language in Science. Still, an Italian edition might be a future project.

Science as Literary Protagonist

This year marks the centenary of Primo Levi, Italian writer, and newspapers, like *La Stampa* of Torino, explain that his steadily growing fame is justified by his stature as a writer, besides his historical Gandhian role as a man, who survived the Holocaust and defeated it by sheer moral strength and reason.

Levi is a Chemist, and, in the *Periodic Table*, the elements, with the memories they trigger, become protagonists breaking the barrier between Science and Literature. The *Table* is hailed as "the finest book on Science", not professorial, yet not banalized, but written with the respect that love commands.

The elements are the foundation of the World, they are produced by and producers of the Big Bang. Many billion years later, the surge of *Life*, its rule, *Evolution* and its peak, *Memory*, are the next possible literary protagonists, indirectly legitimized by Primo Levi.

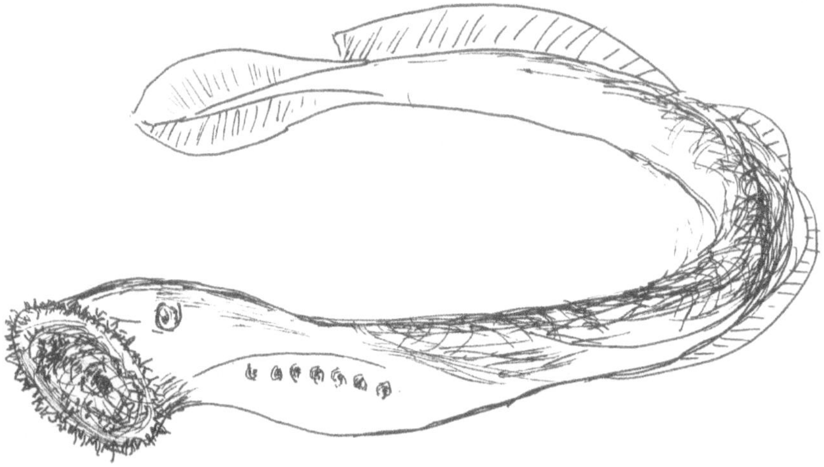

Fig.1 - The *Lamprey*, a Cinderella among fish, even missing the mandibular articulation, is taken to represent, colloquially, a momentous revolution. This early vertebrate sports *Lymphocytes*, which enable the *Specific Immune Response* and the *Immune Memory*.

1

Porcini

A passion for Porcini mushrooms is deeply rooted in my family, descending from both paternal and maternal ancestors. This is not unexpected in Italy, and especially in Liguria and Piedmont, from where my four grandparents stem. When passion meets wit, anything can happen. Giuseppe Giulio Bonino was my mother's father. He liked to be called Giulio, but conserved on stationary, cufflinks and tobacco snuffing box his inscribed initials as J.J. (José Julio, pronounced in Spanish Hota Hota), a reminder that his real life had been in Lima, Peru, where he worked forty years and built a fortune, before settling in Genoa at the turn of the Twentieth Century. In 1929 his youngest daughter Mimma married Guido and from Genoa they moved to Milan. However, every summer she rejoined her large family for a long *villeggiatura* ("vacation"), wherever they chose to go, always welcome, the more so when *nipotini* ("grandchildren") started to appear. For several years in a row the place of choice was Arquata Scrivia, just over the Apennines, a restful and pleasant village amidst hills covered with chestnut trees.

Arquata is halfway to Milan, from where my father comes every weekend, by train. This particular Friday is mid-September and the year one of the very early Thirties. The train screeches and stops, blowing steam. Guido steps down from the tall car, smell of coal smoke, and looks around: Mimma may be waiting for him. Unexpectedly it is J.J. he sees in front of the station, leaning on his cane, tilted fedora, short in stature but strongly built.

"Ciao, Guido! I am here to pick you up: we are in a wave of mushrooms, and…"

Guido is flabbergasted:

"I thought you might like to take the detour through the woods on the way home, we may find some…"

Guido could not be happier - Porcini! and what thoughtful father-in-law he has! The two men take the path and soon are in the wild.

Fig.2 - Guido and JJ seek Porcini in Piedmont

J.J. is professional:

"To cover the terrain, we better proceed in parallel lines: I'll take to the left, you to the right, going uphill. If you find anything, yell!".

It makes sense and Guido, his expectation growing, breaks a forked stick from a bush and begins his walk, slow, tortuous, seemingly hesitating, typical of a seeker. No mushrooms for a while. Then a yell from J.J.:

"Here it is !!"

Guido, excited, crosses to the left and sees J.J. just rising from the ground with a gorgeous Porcino in his hand. Guido is halfway back to his side, hopeful, when he is called back by another shout. He runs, and finds J.J. crouched and about to pick a bunch of three mushrooms from the foliage.

"Oh, Papa! How lucky you are! Would you allow me to do the picking?"

I can guarantee, sticking the fingers in the moss around and under the plump stem and making it pop up free is the peak of the pleasure for a seeker. But J.J. answers:

"Sorry, my son, I can't! It is stronger than me: I guess I like the picking as much as you do!"

Guido shakes his head and goes back to his side of the hill. There will be another three or four shouts, all from J.J. When they arrive home the porcini are weighed: two kilograms, sharp. Mischievous J.J.:

"How unlucky, Guido! I feel really sorry for you! How could they grow all on the left side of the hill?"

Nobody else in the family seems to share his surprise. As it will become clear at dinner, when the mushrooms are proudly served on top of homemade linguine, everybody knew that the devilish J.J. had played another practical joke. He had bought the porcini at the market and, on his way to the station, had planted them on the left side of the hill. Guido (from whom I heard the story years later) had no choice but to join the laughter and the admiration.

This family story is meant as an introduction to something that happened in Sweden, in the second year of the seven I spent there. We arrived in June 1963. The account begins in August of the next year for a simple reason: to my mind, it was inconceivable that the majestic Swedish woods could produce Porcini mushrooms. They were simply too far North, from the habitats I knew. Researchers should know better than to harbor prejudice, but they do it rather often, even in their own field. A typical case was that of the pathologists of Oak Ridge, who could not conceive that rat bone marrow cells may live and proliferate when injected in a lethally irradiated mouse and thus be instrumental in saving the latter's life. Also, immunologist Gerald Edelman, who had shared the Nobel prize for discovering the antibody structure, could not accept the evidence that his pet molecule was flexible. I was there, at a Nobel Symposium in Stockholm, when Roger Valentine's slide, with combinations of antibodies in linear, triangular and square groups filled the screen, their individual shapes adapting to the binding opportunities. Edelman shouted:

"Impossible! There must be a mistake!"

The audience was shocked, and Edelman, who was secretly hoping to receive a second Nobel Prize for the complete amino-acid sequence of the antibody, must have bitten his tongue: too late, no prize materialized. My intellectual block about porcini was a very minor fault, but of the same nature of the examples above.

It was Sunday, noon. Hans Wigzell and I were going to swim in one of the hundreds tiny lakes punctuating the woodland between Stockholm and Uppsala. I was following Hans, when he saw something in the grass, and was about to kick it:

"There are people who eat these..."

But before he could finish the phrase or the kick, I had jumped at him and wrestled him away.

"Are you crazy? These are porcini! PORCINI IN SWEDEN!" I shouted. Then I kneeled in the grass in admiration of the little mushroom family that had cured my prejudice. There were four, on the border of the path, where the terrain was gently descending towards the water. Their brown hemispheric heads formed a perfect horizontal line, because the first in line was the tallest and the other ones were progressively smaller. When I picked these four, a fifth member of the group, tiny and completely white, was revealed, continuing the line like a little soldier, but not yet born; being still underground, sun rays had not touched it, and had not started the tanning process on its head.

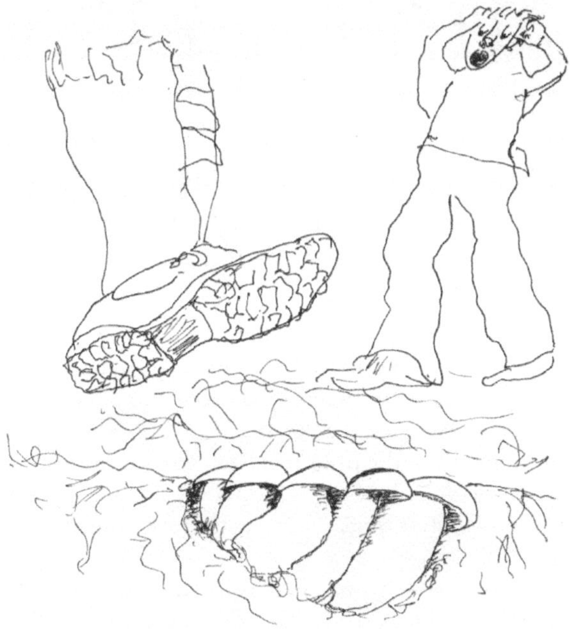

Fig.3.- Franco saves Porcini from being crushed

After the excitement, I was puzzled: "Hans, since they exist here, why are they not famous? Why don't you eat them, instead of kicking them?" He made a grimace, shaking his head:

"Nobody thinks they are particularly good, even when boiled with onions…"

"BOILED WITH ONIONS! No wonder! What fine food could stand such a treatment?"

I became solemn:

"Candidate Hans Wigzell: I consider it my mission to show you the light you missed until this day, and convert you to the porcini culture! To this end, I will drive home, where I have garlic and what remains of the olive oil we smuggled from San Gimignano, in Tuscany.[2] I will join you at your apartment in one hour."

I made two adepts that night: the future director of the Karolinska Institute and his newlywed wife Kerstin, a brilliant, unconventional student who would become a powerful administrator of social affairs in the social-democratic government.

Hans and I worked in the same laboratory at *Karolinska Institutet*[3], and we started a collaboration to test whether antibodies recognizing

---

2. The sixty, or so, olive trees on top of the *Collemucioli* hill have an unobstructed view of the twelfth century towers of San Gimignano, four kilometers down the Val D'Elsa valley on the way to Siena. The production of olives varies from 1 to 6 (or 0) quintals per year, yielding 20 to 25Kg of delicious oil; the hue of the opaque liquid dripping from the press may vary from frank green to golden. That year it had been particularly bright. You can only imagine what it tastes like. No wonder, the Celada family brings it along wherever they go, as this tale from remote Sweden testifies.

3. One of the first questions to my new spirited colleagues, when I arrived at *Karolinska Institutet* in March of 1964, concerned the origin of the name of this majestic Medical Center. It must have been Hans Wigzell who answered: "I know that, for a visitor, the adjective "Karolinska" sounds funny! It refers to Karl Den Tolfte, a.k.a. King Charles XII. This king lived at the juncture 17th and 18th Century, and died at age 32 in 1718. Sweden was not at all the peaceful and peace-loving land that I hope you will enjoy: on the contrary, string of leaders were

sheep red cells and those recognizing horse red cells were made by the same clone of lymphocytes. The project was going fast and the results syndicated that the two antibodies were made by separate clones, supporting the idea that *immune specificity* is determined genetically, and not acquired or "modeled" from the *Antigen*[4]: in other words it follows Darwin, not Lamarck.[5]

---

belligerent, and Charles XII was a real warmonger. He was surnamed "*The Meteor*" for the speed in which he moved in Scandinavia and Northern Europe with a forty-thousand-men army, twenty thousand of which mounted. He put a Swede on the throne of Poland and invaded Russia, can you imagine that! He waited the winter, because the ice allowed his army to cross river Vistula's mound. He must have felt invincible, seeing the Russians flee before him. But the Russian Commander was no moron: his was Piotr, The Great. His retreat was calculated, and left burnt land behind (Did Hitler expect less?). Charles was finally beaten, half his army destroyed, half reduced to surrender. The King with 500 men found inglorious shelter with the Turks, but, when he returned home five years later, he found a mass of veterans, still alive, in incredibly poor health conditions, and..". "Are you telling me that the king built the Hospital for his veterans?" "Well he certainly meant to, and promised, but he had no time to do it, since he had another couple of wars on the burner, with Danes and Norwegians, I'm not sure in which order. Finally, he was shot with a large caliber gun, during his siege of the Danish fortress of Fredriksten. But the needy soldiers remained, and the actual Hospital with Karol's name was built on Kungsholmen in 1810. Hopefully Karol could rest in peace.

4. A toxin or a foreign substance is an *Antigen* if an antibody is able to bind it. Not all antigens can induce an immune response, but those who do are called *Immunogens*.

5. The specificity of the adaptive immune response relies on the diversity of antibodies, so large as to "cover" any, never contacted potential antigen. Their first interaction is totally aleatory, but after that, as "recognizers" of a possible enemy, the cells will be allowed to multiply and become part of the Clonal Selection, the routine proposed in the Fifties. The contenders of the 19th Century, Lamarck and Darwin, were long gone, but all their followers clashed with gusto on the specific response: for the Lamarckian and many creationists the specificity would be "molded one by one" on the antigen, whereas the majority adopted the theories on the genetic nature of Biology, the mutation increasing diversity and the selection of fitness at all levels. (Match score: Darwin-Lamarck: 2-0)

Porcini had nothing to do with the antibodies, but they certainly did something good to our friendship. But, beware: this kind of mushrooms is highly emotional stuff and they may even risk breaking relationships. That Sunday, when I arrived home, I had to make a phone call to Nanni, a biochemist from Trieste, invited for a research period at the Institute. He was also a nobleman, count Sottocasa, a clarinet player, a high-spirited fellow and the husband of Gianna, who had become Antonella's best friend. For months Gianna had been the most beloved occasional babysitter for our children, but recently the couple had moved to Lidingö, an island facing the Stockholm waterway, residential along its rim yet still untouched in its wood-and-meadow center. We kept in touch by phone and frequent visits.

Nanni's passion for Porcini was as great as anybody's, but he shared my disbelief about their existence in Scandinavia.

"Franco! What's up? It's midnight, we were almost asleep!".

"Sorry, Nanni, I could not help telling you: PORCINI GROW IN SWEDEN"!

"Impossible! You must be hallucinating!"

I was prepared for this kind of reaction:

"Well, not only have I seen them: I have picked them. I have cooked them *al funghetto*, I have eaten them, and I have witnesses!"

After a long silence he thanked me: I had changed his summer, and, as a porcini-picker, I had scored high, and my friend's disbelief increased the stake. I wanted to show him the specimen and look at the expression on his face. Unfortunately, I had to postpone that pleasure.

The next day, Monday, I had planned a long experiment on *Immunological Memory*, the project I had chosen for the new lab in Sweden.

The effects of memory were discovered in mid-nineteenth century, when immunity from smallpox was obtained by inoculating a related virus (the cowpox). This procedure, called vaccination by extension, even when the antigen used has nothing to do with cow or cowpox, was first tried successfully by Dr. Jenner and applied to other infections by Louis Pasteur, and has become the greatest contribution of immunology to medicine. However, the mechanisms that make vaccination so precious have been clarified only in the 1960's. Which are the cells that make the defense? *Lymphocytes*, of course. But there are two families: one makes antibodies, which go around and bind to bacteria or viruses in a *specific* way. The cells of the other family are also *specific* but act directly, going around in case of a viral infection, binding to infected cells and killing them before the virus can come out of them and expand the infection. Once a response is successful, it eliminates the foreign invader. But, the following year, the infection can reappear. The organism must immediately verify if the new "invader" is unknown or the same that was previously defeated. In the latter case, memory will tell and organize a faster, better and stronger response, called *Secondary*. This is what happens, but how? My experiments are all directed at answering these questions.

That day, I would be testing whether the immunological memory (IM) decays in time and, if so, at what rate. I collect the spleens of six mice immunized with human albumin four weeks ago, make a suspension of the cells by teasing the tissue with two tiny forceps designed for eye surgery, technically and a bit ironically called *mouse-toothed* The delicate dance of the two forceps on the soft spleen tissue submerged in the magenta solution produces little clouds of free cells and separates them from the stroma, a white filamentous material. With this method each spleen yields about 160 million live cells, twice as many as can be obtained with the usual, less refined methods. I filter the suspension through a fine steel mesh, and hand the single cell suspension on to Birgitta Ashio[6] for counting and adjusting the concentration to 50 million cells per ml. She gives me the loaded 1-ml syringe (the type, thin and long like pencils designed for subcutaneous delivery of insulin) and I proceed to inject the mice intravenously with 0.2 ml (that is 10 million cells) per mouse. The recipient mice are genetically identical to the spleen donors, thus no rejection is expected, but Birgitta has irradiated them, to make sure that the antibodies we were planning to measure during the next month will be produced exclusively by the donor cells. The injection in itself looks like a heroic performance, but I learned to master it during a two-year training at the best of schools: the lab of Takashi Makinodan (Japanese from Hawaii) at the National Laboratory of Oak Ridge, Tennessee. From there I brought the technique to Sweden together with the "mouse-holder", an advanced system, developed and built in the laboratory's workshop to my specifications in order to increase the precision of the experiment and render the procedure more humane.

Since donor mice were immunized with bovine albumin[7] their spleens harbor anti-albumin memory cells, and now all my recipient mice contain an identical amount of these cells. Today is day zero. Tomorrow, day one, I will inject ten of the recipients with albumin, and during the next ten days, monitor the response (technically, the

---

6. Birgitta was perhaps my best technician ever, tall, frizzy, always smiling; her family belonged to a rural, religious minority in the hills of central Sweden. She did not correspond to the typical Swedish stereotype. She did not accept vaccinations for herself, but she was glad to help me actively improve other people's health. After I left Sweden, Birgitta began working on HIV, then little by little, resumed her university studies and eventually obtained a professorship in Norway.

7. The first time the antigen is inoculated is called *Priming*, whereas *Challenge* is the successive infusion of antigen, which elicits a secondary response.

Secondary Adoptive Response[8]) by collecting blood samples and measuring the antibody concentration. The peak titer[9] is the best measure of the elusive memory. The challenge of other 10 mice of the donor group will take place on day 10. Then, on day 20, 30, 50, 80, etc. I hope to measure the dynamics of memory over this relatively short (but still heroic) time, and extrapolate it to the mice's lifetime and further. I am excited. The experiment takes the entire day. I arrive home at 7:30 p.m.

I have just finished the soup at eight, when the phone rings. Nanni is excited:

"Guess what, you were absolutely right! We collected a basketful of Porcini this afternoon, here in Lidingö, in the woods behind our house!"

Yes, he was excited, but his voice revealed some uneasiness. I was hurt:

"What? You went out for mushrooms without even telling me?!"

There is a unwritten law that all Porcini hunters must obey: if somebody gives you a tip, you MUST invite him to the next hunt. Everybody knows this; Nanni knew it very well. That explained his subdued tone.

"You are right. But you had your experiment… as a matter of fact Gianna had told me we should have waited for you…".

I said, coolly:

"Ok, Ok, I understand…" But my voice was sending another message, and Nanni wanted to make peace, clear and sincere.

---

8. An *immune response* is called adoptive when, instead of being generated in the animal that is immunized, the immune cells are transferred from an immunized animal to a non-immunized one. Because the number of transferred cells is known, this procedure affords a more accurate measurement of antibody production per cell.
9. Titer is the measure of the concentration of antibodies in a given biological material such as serum.

"Franco! I have a great idea! Let's go to the spot right NOW! There is still some daylight, and I have flashlights. Jump into your car and drive like a madman"! All my objections were turned down and at ten o' clock we were in the woods. There was practically no light: the longest days were in June, and the famous Swedish daylight was declining rapidly. We turned our flashlights on. In the Guinness Book of Records, we would had qualified in the category *Mushroom Hunt by Night.* In one hour we spotted seven Porcini. The last one, large and with a majestic curved stem loomed in the beam from the top of a boulder covered with moss. Even the spotting distance, eight meters, would have earned a Guinness. It was wonderful. And our friendship was more than saved.

Fig. 4 - Franco and Nanni's Porcini search by night in Sweden

2

Boris

If you want to understand *Immunological Memory* and why vaccines work and save lives, even half a century after having been inoculated under one's skin, you will face the question of how the specific information is stored. I asked this question in Sweden, in the Sixties: could it be that the original molecule from the foreign invader, against which the response may be triggered, is conserved inside the body for such a long time? Or is it the *image* of this molecule that happens to reside in the *memory cells* and therefore lasts as long as the life of these cells or of their progeny? Memory cells are the offspring of *naïve cells* (newly born cells from the bone marrow) and mature during the primary response but, instead of entering the battle, they are kept "on the bench" in high numbers just in case the same infection (antigen) comes back months, years or decades later. If so, their response, *secondary* response this time, is quick and extremely efficient. If the antigen is stored in the body, memory could only decrease because the non-living molecules of the foreign invader do not multiply. But if their image is "remembered" by the system, this "memory" could be stable and may even grow in time.

I had been transferring the same amount of memory cells in recipient mice. Thirty mice per group received one million memory cells and were "challenged" with antigen after 1, 10, 20, or 40 days: the strength of the response that I measured by monitoring the antibody five days after the challenge, had been slowly but consistently decreasing, with a half-life of about 45 days. These results seemed to support the direct role of the antigen.

But how could I prove, or disprove it? I would need to trace antigen concentrations as low as one particle per memory cell, and the best methods available were not sensitive enough to detect this. Frustration. Silent walks in the Swedish woods while the evenings grew longer... However, there exists a secret organization (SFWF: *Fellow Scientists Without Frontiers*), whose members always intervene in these moments. One of them was Boris Rotman. He was spending a sabbatical at our laboratory, at the Karolinska Institute. He was a physical-chemist-microbiologist, coming from Chile by way of California. His stop in Sweden was the last one before settling in Providence, Rhode Island. Boris was a specialist in impossibilities. He had just succeeded in one of these: measuring the digestive capacity of a single enzyme molecule (catalytic turnover). Catalysis is what an enzyme does when it comes in close contact with another molecule, i.e. its specific *substrate*[10]: it digests it; in plain words, it breaks it in pieces. Boris had chosen the enzyme that digests lactose (*beta-galactosidase*) and engineered the substrate to become fluorescent if digested. To the substrate he added the enzyme, and sprayed the mixture in tiny droplets on a slide by using a perfume atomizer. The enzyme was in such high dilution as to make it statistically impossible for a droplet to contain more than 1, 2, exceptionally 3 molecules, while the majority contained none.

---

10. *Substrate* is the molecule that an enzyme is going to digest. The natural substrate is, in this case, *Lactose* and the artificial substrates used are molecules containing the same target as lactose plus a color or fluorescent product.

Boris put the slides face down on oil and let the catalysis work for sixteen hours at 37 degrees Celsius, and then measured the green light spots that appeared, like green stars in a dark sky, when illuminated by a dark blue beam under the microscope. The results of countless experiments showed that a single enzyme molecule could split 3000 lactose molecules in one second.

At teatime, during the customary informal get-together at George Klein's institute, I told Boris what I wanted. But it was immediately clear that his one-molecule detector could not be directly applied to my case, because *catalysis* and *antigenicity*[11] were two very different properties of the enzyme: the former needed the molecule intact, while the second would reside in much smaller fragments of the molecule (called *epitopes*), and still carry all the information for the memory. We looked at each other. I ventured:

"Could we use an antibody instead? An antibody has a site complementary to the epitope…".

It has been a typical trick of immunologists: to label antibodies and infer the number of antigens by *counting* bound antibodies, like measuring the bait by counting how many fish were hooked. Boris replied:

"If you a need a precise measurement, you could use, as antigen, one of those enzymes that are inhibited by antibodies on a one to one basis." (Several of these cases had been described: it is believed that the enzyme activity is blocked when the antibody binds the enzyme's *active site*, thereby blocking the access of the substrate).

We seemed to be making some progress, but I wanted more:

"Boris, this would probably work, but we would be subtracting a small effect from a large background: certainly not a precise method…Is there any way to go in the opposite direction and *create*

11. *Antigenicity* is the property of a molecule to be recognized by antibody.

activity by antibody binding? All the measured activity would correspond to antigen bound and would be recorded against zero background! If fluorescence is used, it will shine against a black background!"

Even now, when it is history, this sounds crazy. Boris, typically, was imperturbable. But it was clear that he liked the challenge: he started to mumble to himself:

"Well, there exist many mutant bacteria that produce defective enzymes ... of course, we should look at those where the mutation affects only a single amino acid (these are called point-mutations); we just have to see whether antibodies can cure them… as a matter of fact I also know who has a collection of these point mutant *E. coli*[12]. It would not take much time… my limit of engagement in impossible projects is 30 days. I am leaving next week, but you must be in my lab in Providence by June 1st! Two weeks from now".

Everything conspired. It was mid-May, school was almost over. Antonella and the three children were preparing to fly back to Italy and stay with their grandparents. They gladly accepted the idea of coming to Providence in the middle of June. I would skip Italy, fly to Providence and arrange the family accommodations. Even my Volvo Amazon needed to travel to the States and be sold there, since it had been purchased under a tax-free agreement, and the cost of transport from Stockholm to Newark would be only $100.

When I arrived in Providence in June 1967, Boris' laboratory at Brown University was already set up, antibodies had been prepared and E. coli mutants were in route from California. Unexpectedly, Boris seemed to have some doubt, or at least some need to be reassured about the wisdom of our project. He arranged for us to

12. *Escherichia coli* is the bacterium that make the Beta-Gal enzyme and we all have it in the guts.

meet with a geneticist at Brown, Seymour Lederberg, the little brother of Nobel laureate Joshua Lederberg. On Monday, we knocked on the door. In a flash of memory from adolescence, I saw a scene from a vintage Mickey Mouse cartoon album, when Professor Ecks and Professor Doublecks (two large apes), puzzled by the failure of their magic beam to hypnotize Mickey like it did with Horace Horse, go to consult a higher-ranking Professor, Triplecks (an even larger ape). Of course, they did not know that Mickey had protected his head with a lead pewter pot that looked like a Roman helmet. Triplecks's response was:

"This brain is obviously different! Go back and study the brain, you blockheads!!"

Seymour did not abuse us. He listened amiably and then said:

"The mutant you are after, which is defective but can be re-activated by an antibody, sounds a bit strange but the rule is: *if you can think of it, it exists*. It may be rare, but that just means more tests!".

This reasoning was reminiscent of Saint Anselm's ontological argument for the existence of God but it was encouraging, coming from a good stock of geneticists. However, I doubt that we would have given up even if it were not. The next day, Tuesday, the E. coli strains arrived and we went to work. There were forty-seven mutant bacteria, each with a different defect in the gene coding for the enzyme beta-galactosidase, and thus all unable to break lactose into galactose. We were going to set up ten cultures every day, and after 24 hours, test them: each culture would be divided into three test tubes containing saline, normal rabbit serum or rabbit antibody against the enzyme. After another short incubation, a gross test for enzymatic activity would be done in the same tubes, by injecting substrate that liberates Nitrophenol. Free Nitrophenol is bright yellow.

Wednesday. We are looking at the wooden rack in front of us, holding 30 tubes, 10 rows of cultures, 1 to 10, of 3 tubes each.

We started the test; nothing seemed to happen. After one hour we could see some faint color in some of the cultures. But no difference within rows, which meant no effect of the antibody compared with normal serum.

On Thursday the rack has cultures 11 to 20. Seconds after the substrate is added, tube 3 of the third row (and only that!), lights up under our eyes. In a few minutes, it is intensely yellow. The rack has 29 colorless tubes and one yellow. Remembering it after 43 years I am still shocked. I had done many experiments in my life, and fortunately I had my share of positive, solid results. But this one was special, so unexpected, so fast, so visual. By refining the analysis, it turned out that the antibody raised the enzyme activity 500 to 1000-fold, and only 2 of the 47 enzyme mutants had this property. We realized immediately that the principle unveiled was important: a genetic damage caused by a mutation in the DNA had been repaired at the level of the coded protein, that is the gene's "product".

I said: "Who knows, this cure in the "product" could have important applications".

But Boris doubted that: "I remember my typewriter back in Chile. It had a "mutation" in the character "G", which printed like "C". At the end of each page I had to take the pen and correct all the Gs in the "product'!

The work in Providence was soon completed. Boris and I spent the rest of the summer weekdays writing the results in a manuscript that was to be published in the Proceedings of the National Academy of Science, and the weekends sailing in Narragansett Bay. Our phenomenon (AMEF, Antibody Mediated Enzyme Formation), would engage me for the next ten years, yielding insights and applications in many directions, except one; the paradox of immunological memory for which it had been imagined and created. The reason was simple; the reconstructed enzyme was less stable than the original and would not remain active for the sixteen hours of incubation needed to accumulate enough fluorescence to spot one molecule for the droplet test. Another lesson to be learned: significance and usefulness of discoveries rarely go hand in hand.

Sobering, but without regret. After all, we were scratching the surface of something much deeper.

A reader may wonder how I would spend all these weekends in the legendary Sailing Academy of the Bay. There are seven Yacht Clubs on the Waterfront and each organizes two regattas every Saturday/Sunday for the seven classes of the Narragansett fleet. At that time, the noblest class, the fastest, and last to start, was the Herreshoff-S, designed in 1926, all parts obligatorily original, and sporting the largest, most graceful mainsail from the top curved wooden mast. Sure enough, Boris was the navigator/tactician on one of these jewels.

Back to the reader's question: how could I find the time, with wife and three lively offspring 7, 6, and 4? And, by the way, did they enjoy New England? These are legitimate questions. And they show that you are an attentive reader… As a matter of fact, the Celada family never made it to Providence. In my white Volvo I drove to Boston airport. The Alitalia flight was greeted by a colorful population of Italian-Americans meeting their relatives. As the passengers were wrestling their way through with difficulty, I was trying to spot my people. Suddenly Piero appears, my brother in law, whom I did not expect to be on this plane. He was alone, and his account was both bizarre and reassuring. At the moment of boarding at the (then) tiny Malpensa Airport, Antonella had already shown the passports, her own, (which also included 4-year-old Anna's), and the American passports of the two older children, Luca and Benedetta, who were born in Tennessee. (This difference of nationality within the family had already been cause of envy and frustration to the Italian-born child. Anna, typically, overreacted when asked by relatives and friends:

"Are you also one of the Americans?"

Her answer was hard and defiant:

"No! I was born in Torino: Torino is my America!".

"We have a problem", the airline officer said, "there is only one visa on this passport!"

"Who could have issued one visa to a passport valid for two? This is total nonsense"!

Both Piero and Antonella tried to fight along this line, but to no avail. No boarding. No Boston, no Providence, where I had rented a cottage on the seaside. Fortunately, the grandparents had accompanied the family. They drove them directly to the Dolomites for the rest of the summer.

Is there another lesson here? Probably yes, but I can't figure it out. Piero is my dearest brother-in-law. He says I was more shocked than I cared to admit by the unexpected "no-show" of the family. He says it even affected my sense of orientation:

"I remember that when we left the Boston airport to drive in your Volvo to Providence, you entered the wrong ramp of the Interstate! It took us two hours to realize that we were driving to Maine!"

Admittedly, he had a point.

Relevance of Discovery: can it be applied to other fields?

A beautiful, clear result, pursued with enthusiasm. The enzyme is defective, distorted by a small mutation. To restore it I use an antibody raised to bind an active molecule. If it manages to bind the mutant molecule, the resulting complex will initially be imperfect, and become functional only when the original shape is restored.

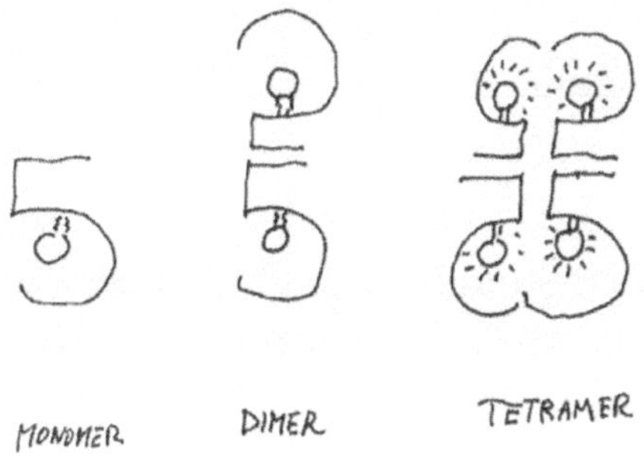

Fig. 5 - The enzyme beta galactosidase breaks the milk sugar lactose into glucose and fructose, making this basic fuel the first source of energy of living beings. The active beta-gal is made of four identical monomers, which combine into a tetramer in a precise symmetry. Each monomer has a "lactose cutter" (pictured as a light bulb in this drawing), but, as shown in the drawing, only in the complete tetramer, all cutters are active (all light bulbs are on). Mutations of certain amino-acids can occur causing a change in the shape of the molecule and therefore the enzyme may not work. In the experiment of 1967 we produced an antibody that binds to the mutant and recovers the active conformation and function.

Two labs will be engaged for ten years to define and completely understand the mechanisms involved. A group in Berlin reads our article, uses the same enzyme and shows that anti-betagalactoside antibodies can protect the enzyme from heat damage. Comparing results, we demonstrate that the two capacities, to activate and to protect, are the deed of binding to different "spots" of the molecule.

Let us go back to our initial stimulus for this research: to decide whether memory depends on the presence of at least one molecule of the antigen (Beta-galactosidase) in each memory cell. That test was not feasible for a technical fatal difference between mutant and wild-type: while a single molecule of the latter stays intact in one droplet and after 16 hours incubation has produced sufficient fluorescence to be measured, the former, initially kept active by the antibody, is less stable, and may end its catalytic work even before it could be measured, making the originally designed test impossible. Pity: our test would have been extremely elegant. Seen many years later, and having continued studying immunological memory until today, I can tell you what the test would have said: Memory consists of selected lymphocytes, which all "recognize" the antigen, even decades after the first stimulation. The antigen will serve as the stimulus to trigger the response; but the presence of the antigen is not required to keep memory alive.

What about the human beta-gal? Soon after the discovery of mutant beta-gal activation, I became aware that a genetically defective enzyme in human infants causes accumulation of lactose in the nervous system and is lethal in a few months. Could our activating antibodies be tested in this very relevant circumstance? Unfortunately, a quick study of the literature showed that the human beta-gal, which causes the disease, is a completely different molecule, smaller and monomeric.[13] There is no possibility at all that

---

13. Monomeric means "made of one piece". A *Polymer* is made by more than one piece. Speaking of proteins, the majority are polymers, composed of 2 (*dimer*), 3 (*trimer*), 4 (*tetramer*) *monomers*. The components are kept together by "week attractions". There may be more than one way in which two identical proteins can stick together, which increases both diversity and functionality. One example is Beta-gal, a tetramer composed, whose monomers become dimers when two of them bind by their head, and the tetramer is formed when two dimers meet "face to face". The shape of number 5 is used in the drawing to facilitate this difficult description: 5's head is flat, its tail is round; there is only one way to put together

the same antibody can cross-react and affect the function of the human enzyme.

In 1969 an interesting researcher/explorer visited *Tumörbiologen* in Stockholm and gave a seminar about Kuru, a mysterious disease that had spread in Papua New Guinea. Dr. Michael Alpers showed us a film, on which he collaborated with Danil Gayduhsek, demonstrating that Kuru is transmitted to the aboriginals, by eating the brain of affected humans, supporting the idea that the pathogen was a "slow virus" transmitted to children and adults in the "ritual cannibalism" taking place after the death of the village elder. The documentary was impressive, as it showed experiments in which primates, injected with the brains of Kuru victims, showed the same symptoms and characteristic "dancing movements" of the sick donors[14].

Time passes. For ten years my laboratory was a "center of molecular conformation", investigating how to modify any protein conformation by using antibodies, which "prefer" one of two possible shapes. The antibody will not impose a shape, but favor one; the rest of the job will be done by the laws of thermodynamics. Master of this philosophy is Roberto Strom; look him up in the chapters about Stockholm and Roma (*Porcini*, Chapter 1, *Blockhead*, Chapter 4).

Time passes: the last years of the Twentieth Century are gone. I am now in America when the shattering news is announced: Kuru is not the deed of a virus, fast or slow. The infectious agent is simply and barely PROTEIN: no trace of DNA or RNA, which would

---

four 5's and obtain a figure with an inside and an outside. Beta-gal works, as enzyme, only as a tetramer; dimers are inactive, the "inside space" is the key.

14. The KURU patients showed a characteristic "tentative" foot movement, that gives the impression of a slow dance.

signal the presence of any bacterium or virus. It is difficult to accept that bare protein can be so dangerous. DNA and RNA are the minds of life, and they are necessary for the reproduction of infectious agents! Yet the infection is transmitted from one person to another, meaning that the infectious agent can multiply. And diseases like Kuru also appear to multiply, even if with different new names, such as Mad Cow Disease, Scrapie in sheep, Creutzfeldt-Jacob Disease, the Amyotrophic Lateral Sclerosis (ALS).[15]

They are degenerative diseases of the tissues protecting the brain and central nervous system. In these diseases the lipoproteins assume conformations incompatible with their function, and their conformation becomes "infective", meaning that the simple contact between molecules causes the transfer of the wrong conformation to neighboring molecules, with a domino effect spreading the malfunctions in the central nervous system. A conformational disease! And I know what "conforming antibodies" are and how to make them. Time has passed again, the last lab was closed in 1996, but this field is still of burning interest, and of great potential medical relevance. For me, a well-placed wishful thinking.

Gliding with Boris

In the last fifty years, I managed to do many things with Boris, besides Science, such as sailing our boats, gliding in Rhode Island and flying single engine planes in Arizona. Our activities were not

15. ALS is the most frequent of this group of diseases, thus the most studied. However, no clear proof has emerged of transmission by Prions, nor for that matter, by any virus.

competitive, however Boris always turned out to be the more expert of the two and a perfect mentor, as he conducted any of the sporting activities with the same seriousness and attention he used in his experiments.

Once he offered a course at Brown University titled *Decision Making in Research and Gliding,* using the art of gliding as a genial metaphor for scientific research. Each of the twelve students was given a private gliding session with Boris in order to understand that managing one's research required the same kind of attitude and behavior than managing one's glide: minimal effort decisions, letting currents guide you, avoiding preconception and external imposition, a real Zen of research.

Another year, he found some seagull eggs in Narragansett Bay and planned to use these birds' flying prowess to help him improve his gliding technique. Possibly inspired by the ethologist Conrad Lawrence, he put the eggs in the lab incubator and managed to be the first living being present at the hatching. The newborns decided he was their parent and followed him everywhere and, after learning to fly, they defended the air surrounding his property on St. James island from other bird trespassers. Paco, as one of the gulls was named, was the best flyer and the strongest bird, who would even fight adult gulls: his "imprinting" with humans had deprived him of the normal respect to elder birds of his species. In the meantime, a flight of glider and Paco was being planned accurately. Boris would drive to the airport with Paco in the car and set him free at the start of the glide. Hopefully, Paco would follow him and cruise around the glider, allowing Boris to locate the updrafts by watching the movements of the bird. Unfortunately, the momentous experiment never took plàce: the day before the scheduled flight, Paco disappeared without a trace. Could the bird's parental "imprinting" have come to fade, allowing him to go back to the evolutionary rules? Boris found it hard to believe that his best pupil had left him and suspected the bird had been shot by a malevolent neighbor.

Boris did not give up gull experiments after Paco's loss. In a new hatch, he managed to imprint four chicks on himself by mimicking four different gull sounds, he had previously heard in the bay, by tapping on each egg and making a specific sound at the moment of hatching. These birds stayed with him for many months, each responding to their individual sound. The most rewarding day for Boris was when George Klein visited him from Sweden and, as usual, wanted to swim. Boris joined him with the four gulls in tow. George was flabbergasted. And what about the gliding project? With these four Boris understood why his project was not sustainable an had to abandon it without regret: he had taken the gulls to the hangar to familiarize them with the glider. But the gulls were scared stiff. This seagull was too big, a monster to run away from!

The next adventure took place in Arizona, when we discovered his plane had a defective tire and Boris decided to change both. You can imagine how complicated the operation we embarked on was. It took us four hours, then we just flew.

Fig. 6 – Paco, the gull

Fig. 7 - Franco and Boris in the two-seater Blanik L13 Glider, made in Czechoslovakia.

Our conversation in the glider:

Boris: "Franco, our tow has pulled us to 1000 meters above sea level. We are unhooked and we are beginning our descent at a rate of about 50 meters per minute. When we reach 250 meters we'll begin the landing maneuver.

Franco: "Oh, I thought we were going to fly around for a while…"

Boris: "This is a glider! No engine, therefore we can only descend".

Franco: "…."

Boris: "But, while descending, we may encounter a column of ascending air, which will take us up at a speed faster than our rate of descending".

Franco: "Wow! Like an elevator? But how will we know?"...

Boris: "Look at the meters: one shows the *Meters per Minute* and the pins indicate the *Up* or *Down* direction… Actually, just as we speak it's changing to UP! We are in luck!"

Franco: "5 meters per minute! It looks good!"

Boris: "Yes, very good! Forget landing for a while, we must work to stay in the elevator. We will circle as tight as possible. The point is to let ourselves be guided by the currents without interfering: in other words, we must be patient and adaptable but always ready to intervene if needed.

Franco: "So as of now the decision-making is in fact a *no decision*!"

Boris: "For the moment, yes! Let's shake on it."

Fig. 8 - Boris Rotman and George Klein swim with the gulls

3

Narragansett

Sailing had been an unfulfilled passion, after a deceptively promising beginning, when three third-graders had found themselves in command of an English Dingy for an entire month in *Sestri Levante*, on the Ligurian Eastern Riviera. That happened just before Italy entered the disastrous war in 1940. "Now is the chance to rekindle the fire", I told myself, when I discovered the protected cove of East Greenwich, the coziest of the seven coves in Narragansett Bay. At the Yacht Club I was filled in on sailing: I learned that each club takes over the organization of the two racing events (Saturday and Sunday) of a weekend; each event consisting of as many races as there are classes. The following weekend, the circus moves to another club, and so on. After 7 weeks and 14 races, the entire cycle is repeated. All this works like clockwork and is considered a tradition. Each event consists of races only between monotype boats, boats of the same design, to make sure that skippers and crews receive full praise (or blame) for their results. This contributes to making the sailing extremely competitive in New England. Any boat type represented by four or more boats forms a "class". At the race the jury call a separate start for each class, starting with the slowest and ending with Class 1, whose boats were designed by Herreshoff in 1924. The Herreshoff-S models are thus the oldest boats, almost an antique, but still the fastest of the Bay.

The racing season was in full fervor. This meant the crews had already been formed, and it was difficult to find openings, especially in the best classes. Also, my references as a visiting scientist did not break any ice in this mariner society. But in Class 3 a crewman had broken his ankle. *Mors Tua, Vita Mea* (latin for "Your death is my life"); I was hired on the spot, and ordered to show up the following Saturday at 9 a.m.

The monotype of Class 3 was the Ensign, a compromise between sports and leisure cruising: a 23-footer, with a relatively large cabin, which in "our" boat was used only to store the sail bags. As another mark of ruggedness, the toilet had not been installed. Seen from windward, when keeling, the Ensign boats were round, like plump sailing tea-pots. The owner/skipper was Captain Forrester, sixty-five, large, sun burnt, mustached, stingy on the quality of beer he provided during the races. For him sailing was the best retirement, with the added advantage of keeping him long days away from Mrs. Forrester, according to sailors' gossip. The other two crew members were longtime friends of the captain, expert, placid, kind and thirsty. In this company I, at 36, looked like a youngster, besides being inexpert. "Cabin boys" are disposable, I had to do well if I wanted to be kept for the summer.

The race started at midday. It was a long race, reaching Newport, rounding some smaller islands and expected to finish between 4 and 5 p.m.

Unfortunately, there was no wind. The sun barely penetrated the mist, the boats moved without breaking the water. After an hour I could not hide my disappointment: Captain predicted that the race would be called off soon…

A slight breath of air was felt. Sails that had been sadly flapping filled up more and more. All four of us had to move to windward for ballast. The wind increased. Water gurgled around the keel. We left a white trace. Things were moving now. At tacking, the jib was ripping my hands like a wild horse.

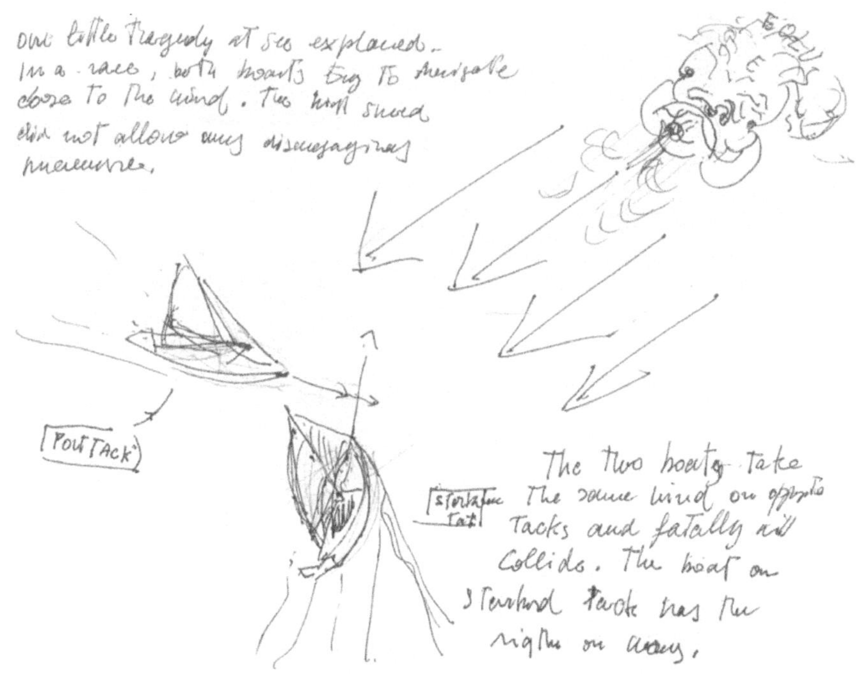

Fig.9 - Currents and winds in Narragansett Bay

We all were excited and focused on our tasks, while the water caressed the deck line at port side. Nobody was keeping an eye on the race. Suddenly I heard one of the crew shouting like a madman:

"STARBOARD! STARBOARD!"

Fig. 10 – Sailboats on risky route

He was addressing another boat, which I saw materializing front-left of us, on a collision route. The shout, translated for the laymen into plain English from the curt seamen's shout, meant:

"Hey, you guys!! You are crossing our way!" To avert collision one of the boats must come about (i.e. turn) in a hurry. And this means YOU, you blockheads, since we are in a starboard tack, and have right of way. Act fast or pay the consequences"!"

If they did understand, they had no time to act and, in a few seconds, we hit them square on the right side. Actually, since they were keeling towards us, our bow hit the board line low and proceeded to mount on top of their deck, with a horrendous noise of broken woods and plastics. Both vessels stopped.

Fig. 11 - Boat collision, detail n.1

"Did we kill somebody?" was the same thought in our four brains.

Fortunately, the other crew reassured us: only material damages, and they admitted their fault. They really had no choice[16].

While these exchanges took place, I was standing in the cockpit, and became aware that my feet were wet. I looked down and saw water, which was ankle-high and rising. I flung open the cabin door. There was a gash of light at the end of the dark tunnel. From the bow, torn in the collision, a large jet of water was springing. I dived without thinking into the half-flooded cabin, and stuck my left elbow in the hole (which was exactly the size of an elbow), while keeping my face squeezed under the deck, to be able to breathe. The water stopped entering.

The crew understood the danger immediately. They knew (I did not) that the Ensign boat did not have any air compartments to keep it afloat in these circumstances and would sink directly to the bottom

16. When they touch, the two boats are on perpendicular routes, both leaning heavily under strong wind. We are on Starboard tack; they are on port tack: they should change direction according to all laws of navigation, but they did not find the time to do it. By now they know that they will be responsible for any related damages. Being on Starboard also means leaning towards port (the left hand if you are on the boat looking towards the bow). In fact, we have to move our weight towards starboard to balance the effect of the wind. The opposite boat (a.k.a "them") presents their starboard side low to the water, and this determines the mechanics of the impact, as shown in the drawings. Our boat mounts over their deck for almost a yard, and the contact takes place well under our waterline level. The collision is a fast process evolving in a matter of a seconds. When our bow touches their board rim, our point is already one yard over their deck by rebound, and due to their speed, the movement is repeated 3 or 4 times. At every contact, the bow cuts further along their board rim, like a saw. The noise of breaking wood and plastic is incredible. All their port cables are severed. It dawns on us that their crew could be hurt. Are they all alive!? But suddenly we also realize that the "assailing" vessel may be the most in danger: our elegant bow-line was not constructed as a saw, or reinforced with steel, like the Swedish ship, MS Stockholm, which sunk the Andrea Doria in 1956. A large hole developed in our bowline, and both racing sailboats are not staying afloat: there is no time to spare!

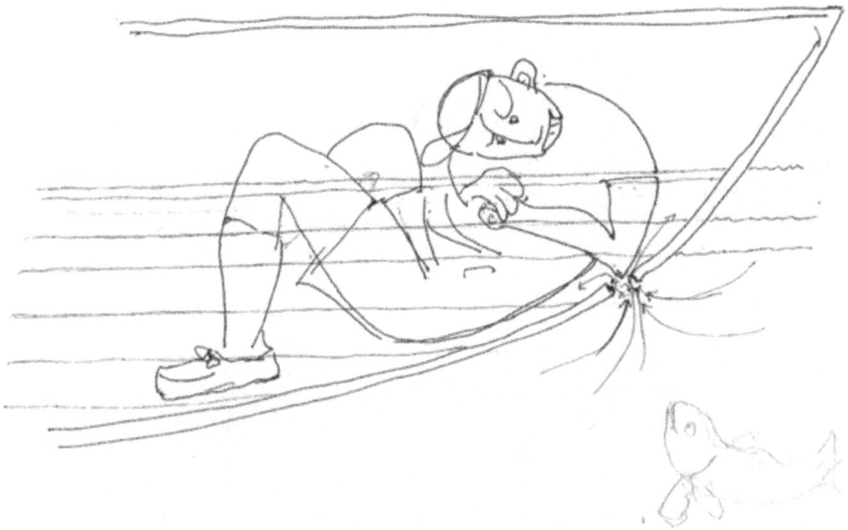

Fig.12 – Franco keeps the boat from sinking

of the Bay. They also realized that my position was precarious: I would have been the first to follow the boat. They looked around for something that could replace my elbow. I heard shouting:

"PANTS!"

A moment later two pairs of jeans were passed down and, in less than one minute, I succeeded to disengage my elbow, receive the full force jet in my face, frantically stuff the pants in the hole, retract from the bow, spit and blow bay water out of my mouth and nose, and finally crawl back to the normal world through a forest of naked hairy legs.

I had saved the boat, and savored my moment of glory: this type of news diffused fast in Narragansett, even decades before the cell phone.

Boris was anticlimactic:

"You did OK - he muttered - but you were not the first one! Remember the little Dutch boy who saved Amsterdam by sticking his finger into the dike?"

Then, wet and shivering as I was, he took me straight to the lab and put me into the large, 37° C incubator room where the cell and bacteria cultures were grown. It was a long shot from a Swedish sauna, which both Boris and I adored, but it did the trick.

Fig.13 – Boat collision, detail n.2

4

Blockhead

I am a good example to illustrate this dictionary entry:

"*Somebody who, as a pure gift, receives a revelation and then forgets all about it, just when it could help him make a wise decision*".

The thin thread of the story begins during my New England "work & discover" summer, in 1967.

In a late summer visit to Juniper Point for a picnic on the lawn, Sylvia Crane, busy with hostess's duties introduces me to an Irish poet:

"Meet Padraic Colum, the friend of James Joyce!", gives us a bright smile and runs to meet some other guest that had just arrived.

Colum is a frail old man, with a winter coat and a felt hat despite the warm weather, a thin pointed nose and tiny vivid eyes. He is 87 and has moved to New England during the last decade. I learn that he was born in Dublin a couple of years before Joyce: the two had grown up in the same literary milieu of the turn of the Century and Joyce's voluntary exile from Ireland did all but decrease their professional and personal interactions, which would last for their entire lifetime.

I just happen to be reading Ulysses - with many interruptions - and I am both in admiration and puzzled by the book and its author. Full of questions, I do not want to miss the opportunity of being answered by someone so close to the horse's mouth. Colum's kind smile encourages me and I hear myself say:

"What a stroke of genius, to recreate the Odyssey as an itinerary through the pubs and flophouses of Dublin! But I really don't understand the choice of the protagonist: what do the two have in common, Odysseus and Leopold Bloom?"

Colum smiles again, and I am afraid he considers me hopelessly naïve. Instead he says:

"Look, having been his friend does not give me the power to understand all the mysteries of his mind. But this time you are in luck, because I happened to ask precisely your same question to Joyce, and I can share with you the explanation he gave me!"

My jaw drops, I am staring; he continues:

"Despite all their differences, both Ulysses and Leopold, find themselves wandering in a perilous world not by their own choice but pushed by Fate, or the Gods, or unknown forces. And as a price to pay for survival, both must sacrifice a piece of their identity: their name".

"You mean, when in the cavern the Cyclops asks: *Who are you…?*"
"Exactly! When Polyphemus asks his name, the hero, to protect himself and the remaining companions, is obliged to hide it, and says: I am "*Oudeis*", which in Greek means "No-man. Thus, he drops the second half of his name, "Seus", which sounds like Zeus, and therefore represents the godly part of Ulysses' identity. On the other hand, Leopold, like most of the Jews of the Diaspora, has sacrificed his original, biblical or sacred name, whatever it may have been, to

survive in mid-Europe, and must wear the almost ridiculous "Bloom", dealt to his family by some Border Officer of Prussia or Poland.

You see: this is what they have in common! It is a very Joycean mystery, based on simple and approximate assonances, but with a touching, human meaning".

I am moved and grateful to Padraic for this "personal communication", almost from the horse's mouth, which I like and cherish. I never saw him again. He died three years after our encounter. For a very long time I could not remember his name, even if his words were imprinted in my memory. Only recently did I look him up in Google: he was, it turns out, quite a well-known writer, author of more than fifty books, among which, to my delight *Our Friend James Joyce*, written together with his wife Mary Colum. The book is out of print. The thoroughly used copy that I obtained from Amazon had belonged to the library of Brown University, and must have honorably discharged after thirty years of loans to students, all colorfully noted on the inside back cover. The book is not scholarly and does not mention the name-hiding story (making Padraic's communication to me even more "personal"). It is an account of professional and personal encounters between the Joyces and the Colums, in Ireland, in Paris (where both lived long stretches of their lives), in London, in America. James comes out vividly, as a difficult genius, feeling persecuted and lonely despite being adored. In particular, the report of Padraic, asked by James to accompany him to see a Dublin publisher, Mr. Maunsel of Maunsel & Co., to discuss the last details about the publication of *Dubliners,* is heart-breaking. Not only did the publisher change his mind about the already signed contract, allegedly for fear of being sued for slander, he also refused to return the manuscript to the author! Colum at this point comes to agree with Joyces' suspicion that strong powers, certainly the Church, and perhaps the Government, were involved in a conspiracy. Later, that ogre publisher would burn the manuscript to preempt a legal action that would oblige him to give it back! The voluntary exile of the author became definitive. If getting rid of him was the purpose of the "powers", they succeeded.

Four years after my New England summer I went back to Italy, accepting a position at the Italian Research Council (CNR), in Rome. I never felt exiled in Sweden, during the seven years I spent there, but when I began looking for positions in Italy (I had this idea that I should prove myself worthy also in my own country and culture), I had the unpleasant feeling of having been singled out as *persona non grata* in the Italian Academia.

Every time I was invited to give a seminar or for an interview in some institute, for instance at the University of Pavia, which seemed willing to develop an Immunology lab, the reception would be enthusiastic, but after ten days I would be notified that the program had been canceled or postponed. Ten days was the time it took the sponsor to contact my former director at the University of Turin where I worked from 1961 to 63, and receive his thumb-down.

Roberto Tosi, my friend and younger colleague, who had spent several years in Torino before joining me in Rome, was not surprised at my experiences. Here is his narration:

"On the silver coins of the United Kingdom you can read this Latin inscription: "Nemo Me Impune Lacessit" (Nobody will hurt me and go unpunished). You see Franco, Italian professors are kings, even if downsized, and they cannot stand any action or behavior that would amount to disrespect, an "injury to their majesty" - in Italian, "Lesa Maestà", Lese-Majesty - consequently the perpetrator will never be forgotten nor forgiven. In my case the damning act had been unintentional and quite foolish: coming back from the US, still

a student, I went for the interview, but I forgot to remove a tiny pet squirrel (a farewell present from American friends) from my breast pocket.

The interview went well and I was hired. The rodent was not mentioned. I was sure it had not been noticed. But when four years later I went to see Ceppellini to tell him I had accepted your offer to work in Rome, he was full of rage, and shouted: "Good riddance! I never liked you, anyway, you and your stupid squirrel!"

Fig.14 - Roberto Tosi's interview

Back to my own case, what constituted lese-majesty was my decision to go to Sweden on my own, instead of asking the Professor to "send" me. The blow must have been felt when I told him that I had applied for an *Eleanor Roosevelt Fellowship* (one of the most prestigious at the time), but he restrained himself, and even unwillingly wished me "good luck!" When two months later, it was March, I received the notification in print, he was not in the institute, and I called him home with the news. He said, bitterly,

"Congratulations! And when are you planning to go? In October? End of the Year?"

"Oh, no", I answered, "I will leave in two weeks".

There was a strange noise in the receiver and the line dropped. Later, I heard from Mrs. Ceppellini that, in rage, her husband had broken a rococo side table by hitting it with his fist, and hurt his arm.

Fig.15 - Ruggero Ceppellini breaks a coffee table with his fist

Yes, the diagnosis of Tosi was correct: it was a case of "injured" majesty and not only metaphorically.

My eventual return to Italy did not contradict the unforgiveness of the sin just described: It was the Italian Research Council, and not a university, that accepted me. In fact, besides me, it accepted two other researchers who had done well abroad but were finding difficulties coming back. The Research Council decided to open an Institute in Rome for our sort, but the procedures were surprising. Our names were kept secret and when we were individually summoned for an encounter with the CNR authorities none of the three knew who the other two Section Heads were. The revelation, a real *Coup de Teatre*, took place in the lounge of the Hotel Mediterraneo, near Rome's Central Railway Station, Stazione Termini. We were introduced and shook hands with each other: the other two were biochemist Corrado Baglioni and neurobiologist Rita Levi-Montalcini. I was pleased. Immediately we elected Rita, well known, and who had already a group of pupils ready to join her, as the Institute's director. The name was *Laboratorio di Biologia Cellulare*.

At the inauguration party, champagne-filled glasses clinging, Rita asks:

"I never heard your family name. What is its origin"?

"Celada is quite rare. All I know is, it comes from Spain. There is a place in Asturias called Celada de la Torre".

"Spanish town and river names… certainly Sephardic! Any family memories?".

"No, but it's certainly possible…"

"No, no; forget it! If you haven't heard about it, you are not Jewish!".

In the new laboratory, adjacent to the majestic Piazza del Popolo, I was still toiling with the antibodies that created enzymatic activity, our discovery with Boris Rotman at Brown University, Rhode Island.

Recently in Stockholm, thanks to a visitor from Rome, a substantial progress in understanding the phenomenon had been made. Roberto Strom, curly red-haired with a 180-degree toothy smile, was the youngest member of the physical-chemistry schools that had cracked the hemoglobin puzzle.

We had done many antibody-mediated enzyme activation experiments with Roberto Strom during the summer of 1968 in Sweden, measuring the effect of systematically diluting the antibody or the enzyme. The results were quite interesting: it seemed that to reconstruct each tetramer, one single activating antibody was required. We began to imagine that an antibody, which has two arms, each ending with the *paratope*, the active site that can bind the antigen, could bind two monomers and drag them together. We were unhappy about this idea, because if dragging was the mechanism, then two antibodies should be engaged by each enzyme molecule, and the kinetic data did not support this… The cutthroat experiment had to break the activating antibodies into pieces, each comprising only one paratope.

The technique to produce *monovalent* antibody molecules was a classic since the cracking of the antibody structure and had been devised by Porter on the way to his well-earned Nobel price: digestion by papain, a papaya-derived enzyme. Our "papain-treated" antibodies were able to activate the mutant: the "dragging-together" hypothesis is out. Conforming antibodies were acting in a subtler way. One single, monovalent antibody causes all molecular regions to change their attitude and behave as they should. Wow!

I wanted to go to the bottom of the antibody-mediated activation; to this end the antibody should be not only monovalent, but also

MONOCLONAL. This assures that all antibodies in the reaction tube are absolutely identical, and ready to be analyzed to the utter molecular detail by X-Ray diffraction, a technique that requires gram-size masses of homogeneous molecules to be converted into crystals. This is going to be my program for the laboratory that will open in Rome.

Few days later, a young fellow, almost a boy, comes to the laboratory. The first thing he notices is that people address each other by the informal "tu" and using the first name. This is normal for labs around the world but not yet in Italy, where the head is still called "*Signor Professore*" and addressed by the courteous pronoun Lei. The young fellow likes this openness. He is a fourth-year medical student at La Sapienza University and has read several books of Immunology, in preparation for this interview. He has even read some papers I have published in Sweden, and looks determined:

"I want to work in this lab, and do my thesis work with you!"

I am impressed, but remain cool:

"What are your interests in research? Do you know what monoclonal antibodies are?"

"Of course I know what they are! And I read the Kohler and Milstein article in Nature!"

The boy knows everything. I need him, he will be a good addition to this lab.

"Well, let's begin to write down something". I sit, pick up a pen and look up at him: "Name?"

"Roberto Accolla". Silence. My pen stops in mid-air. Then, I utter:

"Sorry, I cannot take you." The boy opens his eyes wide, and stares at me. I explain:

"You know what happens every time I call "*Roberto!*" through this door? Three people show up: Dr. Strom, Dr. Tosi and Dr. Revoltella! If you come too, there will be four!! Positively crazy! Impossible." Accolla looks at me like a wet puppy.

... "Unless..." he watches me... "unless you change name."

"But..."

"But I am sure you have a second name, we all have more than one name - don't we? - even if they are not as official as the middle name of the Americans: use that! Sergio? Perfect, a nice name too! Then it is settled. Your Thesis will be: *Anti-Beta-Gal Monoclonal Antibodies*. Let us call in the Robertos for the official announcement."

Months pass. The lab is fermenting with activity, as labs do in their first year of life. Sergio Accolla is well integrated, even if it has not been easy for him to convert his mental enthusiasm into the know-how necessary for the hundred tasks and procedures at the bench and in the animal room: pipetting and, before that, make the pipettes like glassblowers do[17], collecting blood from immunized mice, separating the serum from the cells and collecting the antibodies. All the things that, at Sergio's age I had learned, under more pressure, at the Institut Pasteur on Rue du Docteur-Roux in Paris. Sergio has good mentors in the lab. On the antibody front, Dr. Tosi[18], a.k.a. the *Natural*, on the enzyme front Yasna Radoykovitch, a Chilean with Serbian ancestors who came to Rome, after years with an exacting boss: my friend Boris. This group of three engages in the hunt for activating monoclonal antibodies.

---

17. At that time, researchers still made their own glass pipettes by heating up a glass tube over a flame, while rolling it and pulling its extremities to the desired length.

18. As in all laboratories, one of the researchers stands out for the precision and elegance of his routine lab gestures: that was Roberto Tosi in Rome.

Mice are immunized with the intact enzyme and a week later their blood contains lymphocytes ready to produce antibodies. We are interested only in a very rare sub-population of these cells, those producing the antibodies that cause the *activation* of the mutant enzyme. To this end we will use Mother Nature's procedure: grow all and select the few. However, normal lymphocytes do not grow in a tube or an incubator, they die off after few cell divisions. Bacteria grow spontaneously. Tumor cells also grow spontaneously, even those of the same family of the lymphocytes, the *lymphomas*. It makes sense: one of the things we know about tumors is that they have lost the capacity to accept regulation, typical of normal cells, thus becoming "outlaws", free to multiply without limits. This is why working with tumors is easier. This is also why the idea of Kohler and Milstein becomes the key to the monoclonal antibody. Their idea is to "fuse" two cells together: one, a tumor cell, carrying the "*always go*" property, and the other, a lymphocyte, carrying the already triggered antibody-making machinery: their winning bet was that two fused cells will become a hybrid of the two, baptized, in that occasion, *Hybridoma*. By multiplying, the hybridoma then becomes a healthy and active CLONE. In our lab the successive steps are easy. A clone growing larger than 100 cells is well visible on the bottom of the plastic dish under the inverted microscope, and can be pipetted out and cultured by itself in a small cup. The next day Sergio adds a drop of the supernatant[19] of each of a hundred small culture wells to a hundred tubes containing the inactive mutant enzyme. Then he adds a drop of the substrate, that is a colorless compound that, if digested by the enzyme, will turn bright yellow. If one of the tubes turns yellow, we are in luck: it means that

19. In any cell culture, the cells lie on the bottom of the culture dish or well and the liquid part of the culture is called *supernatant*.

the cup corresponding to that tube contains antibodies able to activate mutant, otherwise inactive molecules.

Sergio turns out to be a dedicated researcher, after a year of apprenticeship. I am proud of my first choice in Rome. I am also proud at a different level, for the caper that solved the problem of "homonymy" in the lab. Actually, I remember myself boasting and laughing about the episode, with colleagues and visitors.

Starting tomorrow Sergio will begin the procedure to isolate the clone: a series of cell cultures will be started, each from a *single* cell of the positive *culture well*, and, after a week, they will be re-tested for antibody. Perhaps 10% of the wells will turn out positive, at this stage, meaning they will produce monoclonal antibody, and from now on there is only a clean, patient work of culturing left, in order to obtain a workable number of antibodies.

Then the paper is written. Sergio is the first author, I am the last. Rita Levi Montalcini, a member of the US Academy of Sciences, sponsors this work for publication in the famed Proceedings (PNAS). Sergio defends his thesis successfully (I am his discussant) and becomes a Medical Doctor. Later on, in mid-1975, I am off for a sabbatical year at the University of California. I lose track of my ex-student.

Genoa, ten years later, 1985. I am a professor at this University, but ready to move again to the States. Cafeteria of the Hospital, where researchers have their meals. Dr. Accolla steps in. I have heard he has got an appointment at the Cancer Institute. Funny he did not seek out his former mentor and greet him. He must be busy setting up his lab. Our eyes meet, he is surprised. We greet cordially, but I feel his embarrassment. To break the ice, I say:

"Nice to see you after all these years" ...then, like cracking a joke "You are no longer Sergio, I suppose... let's talk of our times in Rome!" But, out of timing and out of character, he utters coldly:

"Look, let us talk of what you want, programs, experiments, the weather, but please, do not mention anything about my name. I had enough."

In an instant, memories from different epochs hit me and combine with each other. Roberto Accolla, here, proves that Joyce was right: the name of a man is more important than we think, it is more than the *symbol* of his identity and of his conscience. It is a part thereof, a large part. Yes, with the best intentions, yes, with no ill feelings, I had been a cavalier and should have known better; and I am a blockhead, since I had been warned.

The Human Side

Part of the Roman adventure, and perhaps compensating for the blunder of the prohibited name, I remember the five years of my Immunology Unit at the LBC (*Laboratorio di Biologia Cellulare*) as a rich and electric period. Science was combined with the heavy breath of Rome, the past history mixing with the pulse of progressive political movements. All researchers of the staff had come from experiences abroad, mostly in the USA, and brought their young families and their conviction that this time "everything was possible, thus we will succeed"! Yes, this time the start was right. The signal had come from Paris, in May 1968, and had triggered waves reaching all corners of Europe within a decade, with their content of freedom at all costs. In Italy universities were hit first, in 1969, when students occupied the classrooms and changed the rules. National labs, like

Naples' *International Institute of Genetics and Biophysics*, were in revolt in the early 1970's, while LBC was spared thanks to the internal democracy present from the start and a sense of independence from the decadent university system.

In the city all primary and secondary schools were brewing and the turmoil peaked in 1977 with occupations, assemblies, marches, riots followed by brutal police violence. Civil freedom became a recognized ingredient of Science. In our Lab it was Roberto Tosi and his wife Simonetta, who had a strong and positive impact on the work and the energy of the Immunology Group. By refusing to separate his humanitarian and civil motivation from his serology expertise, Roberto used vacation time productively: every Summer he would pack his precious reagents and fly to Havana via Barcelona, to help his Cuban colleagues measuring the compatibility of kidney donors, a must to enable the life-saving transplantations on an Island under Cold War siege.

In the meantime, Simonetta, a gentle mother of two and an uncompromising early feminist, used her MD degree to found and promote the first Women Counseling Center, run and managed by women, and dedicated to diffusing information on family planning and contraception. Her MD title was essential for the Center, and helped breaking the blockade by the Catholic Church, which threatened attending physicians with interdiction from sacraments. The Abortion Law was being discussed in Parliament, and a new law (Law n.194) was finally voted in 1978, and more effort was needed for its implementation. Simonetta was untiring and became a national figure. After her untimely death in 1984, the City of Rome named *Via Simonetta Landucci Tosi* after her.

Not every week did something extraordinary happen, like when news broke that in Santiago the Pinochet-CIA organized coup murdered Salvador Allende.

Fig.16 - Salvador Allende under siege at La Moneda

Nobody worked that day at our Biologia Cellulare lab. The city was stormed, the Police had to defend the US Embassy. However, in normal working days the personal problems of some of the quietest post-doctoral researchers may demand the attention of the more experienced.

Oreste Acuto, Biology student at La Sapienza University was the first to ask to work on his thesis in our lab, before Roberto Accolla.

Oreste chose Tosi as his mentor, for an evident character affinity; his work was excellent and he graduated with honors in November 1973. It was the first graduation of the Immunology Unit, and was duly celebrated in our lab. Our founding rules said that if a graduate wished to continue his/her activity here, he or she should first get a work experience abroad, in a high-raking institution. At first, Oreste did not seem very eager to look around, but by the break of Summer, following a hint by Roberto Strom, he got a one-year work offer in Baltimore. The lab's field was serology, the director, an expatriate Italian who had a grant and was eager to fill the position. It was not difficult to write a strong letter of recommendation for Oreste, which I promptly did and sent. Within ten days Oreste received the contract, beginning July 1, together with a one-way airline ticket to Baltimore. On the morning of his departure we had a party in the lab, even bigger than the graduation party. The girls had wet eyes and brought all sorts of presents that might be useful in the new continent. The present of the lab was a big Webster's American Dictionary of English Language with my dedication. Oreste's girlfriend was missing. It was Monday, July 1,1974. The taxi bound to Fiumicino Airport arrived in Via Romagnosi (the LBC address, two blocks from the incredibly beautiful Piazza del Popolo). After a collective bear hug, Oreste was off. Pass Tuesday. Pass Wednesday. Perhaps we will receive some news about the arrival, by tomorrow... no news on Thursday. On Friday morning Oreste Acuto is back in the lab. The shock of shocks: Oreste comes to my office. I ask:

"Are you ok? Tell me what happened to you!"

"Franco, it is not easy, because there were many things happening together. I have not slept during the flight, and I did not close my eyes until boarding on the return flight… In Baltimore the heat and humidity was unbearable, from the moment of landing, except when it was replaced by the noisy storm of air conditioners…

The car from the airport to downtown crossed overpopulated, poor and dangerous neighborhoods, people looked strange to me, they seemed in pain or under the spell of substances. I would never be able to live there without feeling co-responsible for their misery. I wanted to flee, while all my friends in Rome envied me for my luck…"

"You said that you felt stalled… You had left your girlfriend in Rome. Was the sadness of separation adding to the…"?

"No, no! - said Oreste, lively - She was not the problem! Actually, she was the solution, and I love her for this! The first night I was there we were on the phone when I told her I could not stay but I was trapped. She said calmly: 'Oreste, there is only one way out. You buy another one-way ticket and take the first flight back. All pain will disappear".

"So did I. It took seconds to see she was right".

"Oreste, I don't know that girl. Is her name Arianna?"

"How could you guess"!?"

Life in the lab resumed. After a week Acuto was back at his bench. After ten days a letter from Baltimore was delivered to my desk. The man was furious, and he wanted to hurt. His formal letter ended with the following sentence:

"It is sad to realize that Italian Academia has declined so low as to not being able to recognize mediocrity!"

At first, I wanted to ignore the blow, but I had written the recommendation; I had to answer.

"Dear colleague, I am sorry you witnessed an emotional/existential crisis in a sensitive person. I don't know what happened, but mediocrity is the wrong diagnosis. P.S. I bet that twenty years from now Acuto's scientific production will be better than mine, and perhaps, of yours!"

I got no answer from Baltimore, but I won my bet easily. Oreste tried more trips aboard. The first was to Zurich at the ETH (*Swiss Federal Institute of Technology*) for two years, then to Lausanne at ISREC (*Swiss Institute for Experimental Cancer Research*) for three years followed by Boston, at *Harvard Medical School* for six years (where he co-discovered the T-Cell Antigen Receptor). He then came back to Europe, in Paris, where I met him four years later well positioned at the *Institut Pasteur.* He stayed in Paris for eighteen years and he is presently in Oxford, on his twelfth consecutive year. But sooner or later he will retire and go back… to Paris.

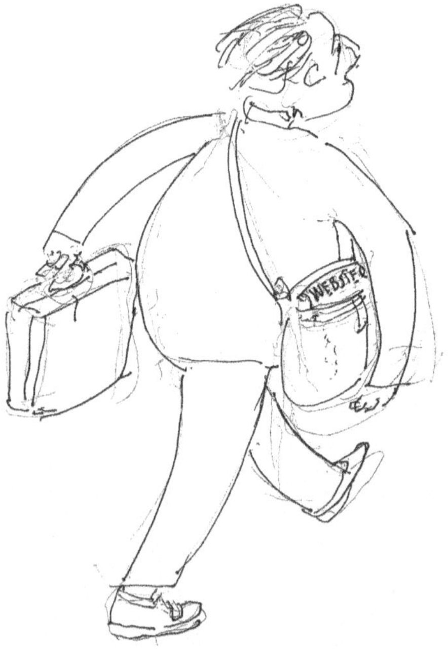

Fig.17 - Oreste Acuto goes to Baltimore

5

Gilbert

Stockholm, late sixties

Gilbert Karpman was the *Attaché Scientifique* at the Embassy of France in Stockholm. He was also a Theoretical Physicist, dreaming of elementary particles and the mystery of time, and I suspect he felt uneasy having traded Science for the glittering but boring world of the embassies. It had seemed to be a clever way to settle his finances and show his in-laws that he was able to provide an adequate social life to their "princess" daughter. However, nothing of his internal debate transpired and he was the most brilliant and efficient science networker I have ever encountered, the quintessential attaché.

We met at the party organized by the *Istituto Italiano di Cultura* (IIC) when I discussed my *Docent*[20] thesis at the Karolinska Institut.

20. The Docent title in Sweden is a degree that certifies excellence in research. It can be earned before or after the license (completion of courses). I already had an MD title from the University of Milan, and becoming Docent made me equivalent to a MD, PhD in the US. The thesis consists of a collection of published papers

The atmosphere was pleasant and relaxed, in sharp contrast to the formality (tails were mandatory) of the dissertation. We sat with our wives around a saffron risotto made and carried hot from home by Antonella, a mountain of fractured Parmigiano and a forest of Chianti bottles, the usual contribution of IIC.

I liked Gilbert immediately and we began to see each other regularly; during our encounters, while chatting, Gilbert was working: he needed to refresh his Biology, and I was a godsend. He had tons of questions and thought my first hand answers and explanations helped him a lot, considering that Sweden was experiencing a new spring in Biology and Immunology. In return, he seemed to find or invent all sorts of connections that might have been helpful for my life of researcher abroad and a bit *deracine*' ("rootless"). These parties in his hands became precious occasions. Typically, whenever a French immunologist happened to travel to Scandinavia, he contacted me and proposed to take him to Stockholm to give a talk in my lab at the Karolinska Institute: he knew that informal (and free!) seminars are the lymph of labs. Or, when Nobel laureate Jacques Monod[21] expressed interest in my new antibodies that were able to modify the conformation of the antigen, (a fresh finding of Summer '67) he mined in the archives and

---

bound together by a summary. The Discussion is led by a competent researcher, in my case the Finn Olli Maekela of the University of Helsinki.

21. Monod and Jacob described the control mechanism of the enzymatic splitting of lactose into glucose and fructose by betagalactosidase, and called it "Allosteric" , i.e. based on changes of shape of small molecules within the so called lactose operon. The enzyme is the same wild-type as the one whose function Rotman measured as a single molecule in a droplet. Rotman and Celada discovered that antibody binding can modify a non-functional betagalactoside molecule and put it together as a functional tetramer. In 1970-71 my stage in Monod's lab at the Institut Pasteur yielded a paper in Biochemisty, co-authored by Celada, Ullmann and Monod.

programs of several French Ministries and produced a fellowship tailored for me to spend a year at the mythical *Institut Pasteur*. At his home, views on the bay and a grand piano, the ebullient Gilbert was always entertaining; by contrast, Madame Karpman, Annemarie, was reserved and her eyes, deep set and beautiful, bore a shade of sadness. She had studied with Levi Strauss, and his brand of Cultural Anthropology was religion for her. This emerged one night, after a conference that had taken me to Buenos Aires. I described the meats I had seen on the carbons during the *Asado* on the outskirts of the Pampa, which had concluded the scientific event. When I mentioned, as a curiosity, the *Cinciullini*, braided tripes that were sizzling on the grill amongst sausages and short ribs, Annemarie's face discolored. In deep emotion, she uttered:

"That is not possible"!

I could not believe my ears. During the rest of the dinner she tried to explain, her voice still tense: in *Tristes Tropiques* Levi Strauss defines a network of ranks that govern the primitive village. Speaking of food and, in particular, meat, there are noble cuts, semi-noble and non-noble cuts; when it comes to ways of cooking, there is a noble way, which is roasting on open fire, semi-noble ways, like casserole-cooking, and non-noble ways, such as boiling. The problem with the *Cinciullini* was this: tripes are the least noble of meats but in my description they were treated like the noblest filet steaks! This was totally inadmissible, regardless of how delicious the result was.

One day Gilbert called me in the lab.

"Do you know Jean Dausset of *Hopital Saint Louis* in Paris?"

"Of course. As a student I visited his lab, he was busy with leukocyte agglutinins. Turns out they are linked to transplantation antigens. This means kidney rejection may be avoided by choosing compatible donors!"

"Wonderful that you know him so well! He will be here for a couple of days, and you can help me!"

It was June, the time the Nobel committee begins to consider the candidates; the prizes are announced in October, and awarded at the Lucia Festen[22] in December. The Nobel rituals are to Stockholm what the Oscars are to Los Angeles: they electrify the air and regulate time for three months out of the year. Gilbert continued:

"The General wants to have him for a lunch at his home, and invite also the *Secretary of the Committee for Physiology and Medicine*, George Klein[23], but we don't want to make the invitation too formal. Therefore, you are the key of this little diplomatic scheme: you know Dausset, you work in Klein's *Tumörbiologen*...."

"But I ..."

Gilbert overruled my hesitation: "It will be nothing less and nothing more than a nice chat among friends"!

---

22. The *Lucia Festen* is celebrated in Sweden on December 13. She is a Christian martyr, originally from Siracusa, symbol of light because of her name. In a part of the world where life in winter is extremely dim, the Lucia Festen is represented by a little girl with her helpers bearing bright light in their hands. The Nobel prize winners are surprised by the visit of Lucia and her team at their Hotel in Stockholm in the early morning on the day of the Ceremony.

23. George Klein and Eva Klein married in Hungary, and escaped from the Russian-dominated country in the mid-Fifties to Stockholm, where the Swedish Cancer Society built for them an Institute dedicated to Tumor Immunology. Under his leadership numerous researchers from around the world but notably from eastern countries found a stimulating shelter and contributed to founding an important branch of Immunology. George is now 87 and well, writes books, develops scientific theories, but I am afraid he has given up the daily swim in the Mälaren, the waterway that envelops the island of Lidingö in Stockholm. In 1963 I took a train from Leiden to Amsterdam, walked to the hotel where Klein was attending a workshop and in half an hour my stay in Sweden was decided. I thought it would be two years but it lasted seven.

At the lunch, the wine and light cuisine confirmed the Cordon Bleu tradition, and General Puget[24], the Ambassador, wore his decorations. Not only was he Gaullist, he belonged to the small circle of men, whom the commander in chief would deploy, with a moment's notice, wherever *la France a besoin de vous!* ("France needs you"), be it at the head of Sud Aviation or to reinforce the diplomacy at the extreme North of Europe. Dausset was surprised by the attention for him, George Klein was interested in the leukocyte groups, which were the counterpart of his own scientific focus, the mouse *hystocompatibility* antigens; but he also told, in impeccable French, some of his famous anecdotes about life in Budapest when he and his group of fellow boy-scouts went underground to escape, first the Nazis and then the Russians. Gilbert reaped a big success in the eyes of the General, while I enjoyed just being there. Was history made, or shattered during that lunch? Certainly not. Dausset did eventually get his well-earned Prize, but more than a decade later, in 1980[25]. By then, most intricacies of the genetics of the *Human*

---

24. Puget, a former *Pilote de Bombardier,* was deployed as president directeur general at Sud Aviation where the French/British Concorde plane was built. Unexpectedly the need erupted to find a shelter for a *Prefect de Police* who was under fire for a blunder in the case of Ben Barka, the Algerian nationalist killed by the secret service. Sud Aviation in provincial, out of the way Toulouse was the perfect spot. De Gaulle called Puget and told him he should strengthen the French diplomacy in Sweden. Puget, obedient, became Ambassador on the spot and, faithful to his self-definition: "*Je suis General, Catholique et Breton, dans cet ordre*" (I am a General, a Catholic and a Breton, in this order) took his work as a mission. One notable action in his new role (as rumor has it) was a victorious coup against a nightclub that shared a wall with the French Embassy. It consisted in eradicating a dispenser of prophylactics that the nightclub had displayed, with Scandinavian candor, on that same wall.

25. The 1980 Nobel Price was shared by Dausset, Benacerraf and Snell, and as it often happens, generated some grudge and discontent among colleagues and competitors. Nobody really disagreed with the choice of Dausset and Snell, one sporting his early intuition of the importance of leukocyte surface antigens (while other colleagues of quality, like Maxime Seligmann were lost in a maze of

Leucocyte Antigens (HLA) had been sorted out, thanks to a worldwide cooperative effort, and most major Medical Centers were building their Kidney Transplantation Services. I had just landed in my next station, Genoa, as Professor of Immunology, after Oak Ridge, Torino, Stockholm, Paris and Rome, and there I was able to set up the essential commodity: a lab equipped for HLA testing, to determine donor-recipient compatibility.

Other embassies also felt the heat of the Nobel fever, when their nationals were involved. One December, the laureates in Medicine were to be three Americans: Luria, Hershey and Delbruck, all founders of *Bacterial Genetics*. But Salvador Luria was an Italian, born and already active in Torino, who, like so many scientists of his generation was mistreated by our dear motherland: Mussolini's Racial Laws of 1938 obliged him to flee, first to France, then to the United States. Luria's presence in Stockholm, to receive the medal from the hands of the young king, posed a delicate dilemma to the Italian Ambassador: how to show post-fascist Italy's appreciation for its offspring, without offending him or the Americans by intruding in their celebrations? It was my friend Sergio Ponzanelli[26], the

---

cytoplasmic structures), the other laying the fundaments of mouse immunogenetics in a secluded island of Maine. There was, however a diffused feeling that the 50-lab international cooperation, led by Ceppellini in Torino and Van Rood in Amsterdam, should have also been rewarded, having allowed to understand the *Allelic Transmission* and the *Supergene* structure called *Major Histocompatibility Complex*. Ruggero Ceppellini was particularly vocal and distasteful in his personal attacks; but these fights certainly question whether the prize to single discoverers still allows an objective recognition of merit in modern Science.

26. Sergio Ponzanelli was a quiet, competent and pleasant person with some fire hidden inside: anti-formalism, anti-bureaucracy, with a shade of anarchism from his youth. He defended two students from Bologna and heckled ("venduto!") the pompous writer Bevilacqua during his speech at the Istituto di Cultura. I arrived late that evening and was appalled. One of the students was practicing in my lab, and is now professor Claudio Franceschi, who brilliantly produces science in the university and popularizes it on TV. Sergio was also instrumental in bringing Dario Fo and Franca Rame to perform in a major Stockholm theater. On the

director of I.I.C, who devised the solution: it involved, once again, a meal. The second conundrum was the invitation: the Ambassador could not afford a rejection. Thus, I was asked to "pull the chestnuts out of the fire" (in Italian: *togliere le castagne dal fuoco*), once again. I had never met Luria, but, by pure chance, in my first Swedish publications I had applied a mathematical formula devised by Delbruck and Luria to test whether a cell distribution is attributable to clonal proliferation. I guess this constituted a sufficient title, and I dialed the Grand Hotel:

"Professor Luria? I am calling to find out whether you are still angry or you would accept an informal dinner invitation extended by the Italian Ambassador?"

Luria laughed:

"Me, angry? I will be delighted! And if it comes to fascists, I fear them more in Massachusetts!"

At dinner a crisis erupted when, seating the guests around a 24-seat round table, the Ambassador's wife saw with horror that two gentlemen were sitting on adjacent chairs while at the opposite pole of the table two ladies were next to each other.

---

opening night, however, no Italian authorities showed up, because the two actors were considered leftist (which they were) and dangerous (which they were not). Only Ponzanelli was there to represent the Italian Establishment. Fo was furious: "Where is the Ambassador? Doesn't' he know that our stay here is paid by the Ministry of Foreign Affairs?" I have lost track of Sergio and his lovely wife Midili' after he retired, but I am sure he rejoiced at the news Fo eventually won, against all odds, the Nobel Prize for Literature.

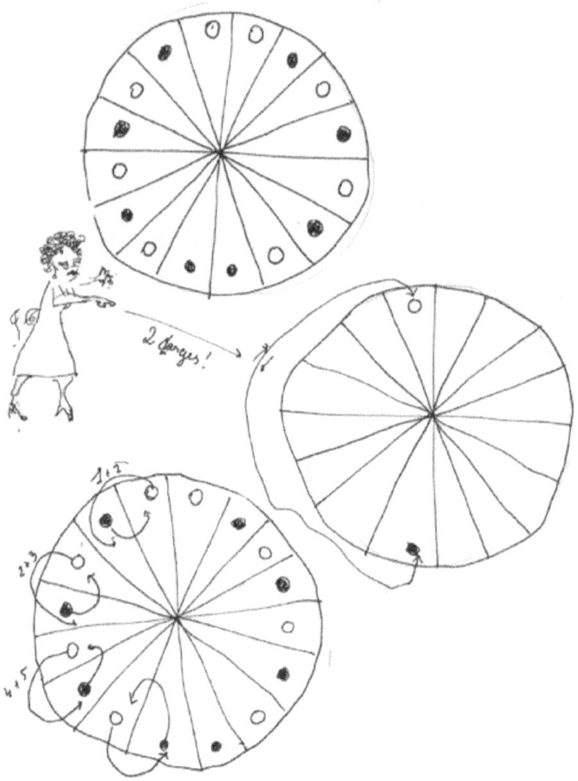

Fig. 18 – The riddle of seat assignment at the Italian Embassy

She realized that this story would be recorded in the annals of diplomacy and, in desperation, tried to solve the problem by asking one of the two gentlemen and one of the two ladies to exchange their seats[27]. This failed, of course, in the midst of murmurs and suffocated

27. This cartoon illustrates the guest seating options around the 24-seat round table of the Italian Embassy. On the left: the distribution that caused drama among diplomats. In the center: the effect of the two-changes solution tried by Mrs. Ambassador. On the Right: the only possible solution, where the "moves" are 8.

giggles. But the guests were mostly scientists who, like the guest of honor, couldn't care less about this kind of symmetry: thus, the dinner was saved. The Ambassador, who belonged to the Turin aristocracy, and the scientist ended up conversing about time past in Piedmontese dialect. The Barolo of the Einaudi[28] estate was conducive.

"Embassies!", mumbled Gilbert when I told him about that evening.

"You are lucky! As an outsider you only see the funny side of it, but I have to live with it."

"Gilbert! What could be so bad? Let the cat out of the bag! (In Italian we say: *Sputa il rospo*!, which means: "Spit out the toad").

"Ok, if you want to know, the curse of the diplomat is to meet other diplomats, always and compulsorily. And in capitals like this one, a routine forms whereby every week an embassy invites the employees of all friendly embassies for a party. Black tie, wives dressed up, wines. It is considered good manners, and a good way to keep informed. Except that, in absence of good old spying, most of the information remains gossip".

"If it bothers you, why do you go?" I argued.

Gilbert considered for a moment my monumental naivete', then explained:

"You see, there is a built-in catch, self-fostering, and deadly. The invitations are handed out by each Ambassador to their own team. If I don't show up, the rest of the circus will assume I was not invited, and that gives them a way of measuring my standing in my own

---

28. The Einaudi family is extremely important in Italy. Luigi Einaudi was President of the young Italian Republic. Giulio Einaudi was the founder of one the major Italian Publishing Firms. The family is also one of the major producers of Barolo, their estate being in the specific region of the Piedmont Hills.

structure. This proves the power of gossip. Of course, I am free not to go. If I don't go once, nobody will notice. But if it happens more than occasionally, in the eyes of my peers in the sister structures and the science offices of the ministries I will be 'the Frenchman on his way out'. As it happens, open horizontal relations and exchange of news and novelties are essential for my work. But who of my *friends* in the circus would be eager to share a good tip with a man 'on his way out' in his own office and in the mind of his own boss? So, everybody goes to the circus, always: the risk, eventually, isn't to miss dances and gossip, but to lose your job."

I was shocked: my lighthearted challenge had elicited a "toad" of unsuspected size. Perhaps I should have restrained myself in the use of Italian folk proverbs.[29]

It was Spring again, the year was 1968 and the month was May. Gilbert was summoned to Paris, to participate to a briefing tour organized by the ministry that would take all French Attaches Scientifiques to visit laboratories and pharmaceutical companies. He thought it would take a week. But that week the Student Revolution burst at the *Sorbonne* and spread like wild fire. Being on an official mission did not prevent the Attaches from feeling the heat. From Stockholm we followed the news with great attention, but it was like looking through a telescope from the wrong end: the happenings were light-years away. The students at Karolinska were busy preparing the spring exams. They did not have time to watch television, with the exception of an occasional tennis match of the emerging junior, Bjorn Borg. Two weeks passed, no trace of Gilbert.

---

29. "Spit out the toad". It is unpleasant to have a toad in your mouth, but spitting it out (recommended) is not without problems: For one, everybody around will know what you were hiding inside.

Annemarie, who had stayed home, did not know what was keeping him abroad. After three weeks she called Antonella:

"Gilbert arrives tomorrow, and he is eager to see you two. He didn't say why".

"Could you come over tomorrow night?"

On arrival, we saw the grand piano covered with freshly collected, (now famous) street posters and handouts with slogans like: *"Faite l'amour, pas la guerre!", "On va vous detruire avec un éclat de rire!", "Il est interdit d'interdire!", "Sous le pave', la plage!"* ("Make Love, Not War", "We will Destroy you with Laughter", "It is forbidden to Forbid", "The sand is on the Streets!") and others, even more abrasive.

Gilbert was changed, excited, his face sunburnt, his voice sounding raspy as though he had shouted and chanted for days with the rioters of the Sorbonne. He had been caught by surprise, but had immediately felt like a fish in the pond, even if he could not actually rally with the students or throw cobble stones at the police. He heard Cohn-Bendit, and collected posters, while the crowd (la foule) rallied around town all day, then reconvened at night at *Les Halles*, at the condemned cast-iron structure of the famous market. There exploded dances and songs, the girls laughed and had given up wearing bras. By pure coincidence, continuous happenings and performances added to their joy: the Italian art director Luca Ronconi was staging the *Orlando Furioso*, a theatrical revolution in its own right. Les Halles were full of whales, flying hippogryphs, and women in love and despair. The knights of Charlemagne were fighting the moors on wheeled horses propelled dangerously in the thick of the crowd: and the people had to make instant decisions, whether to listen to Ariosto's octaves or to run for dear lives.[30]

---

30. The street version of Ronconi's Orlando Furioso is vivid in the memory of anybody who had the fortune to see it (I saw part of it in Milan's *Piazza Duomo*, and took the whole family to see it in the *Blue Hall* of the Townhall in Stockholm).

"You know, Franco, I breathed Liberty after many years, this is the best of France, a new *Bastille*, a new *Commune*. And I am convinced that all our lives, including work, University and Ministries, can be changed and saved from within".

He poured champagne in our glasses and shouted:

"*A la France Libre!*" ("To freedom in France!").

Annemarie stared in awe.

"Tomorrow I am going to see the General; I am sure he will appreciate how the new enthusiasm will improve the spirit and quality of our work. I will offer to speak to my colleagues, and also at the circus".

I sympathized with my friend: I shared his feelings, but I had doubts about his chances to influence the diplomats. Antonella and I walked home with mixed feelings of admiration and fear for Gilbert.

There followed another week of silence. When I saw him again, I did not have to ask him about the meeting with his boss: his eyes sufficed. The General had reacted like De Gaulle in Paris: with rage and feeling of *Laesa Maiestas* ("injured sovereignty"). He was horrified by Gilbert's enthusiasm for the revolution:

"Are you a Communist? A traitor? I should have listened when the *Conseiller Culturel* warned me! No? Then take an oath of loyalty to De Gaulle! Here and Now!".

Then he formally prohibited him to share his feelings and ideas with anyone in the organization. He was mad, but also sad, because he liked Gilbert's work and intelligence, and realized that he and the

---

Unfortunately, there is no record of it. There is another version by the same director and company, staged three years in a castle in Umbria and broadcast in Italy as a six-episode TV show. It is quite different, with less movement, more costumes but the same verses. Many excerpts of it are available on You Tube.

embassy were losing this valuable helper and possible future leader. Gilbert had failed to explain the May revolution, and now he felt buried alive and *incommunicado*. He was also sad, because he had disappointed the old man for whom, after all, he felt a streak of affection.[31]

However, his dream to go back to his elementary particles and discover how they navigate in *quantum mechanics*, became irresistible, and would soon break out in the open. His future would bear a glorious return to University, the teaching of Mathematics at the pearl of French Technology, Compiegne, and the Rectorate of a visionary University campus near Belfort in *Franche-Compte'* in Alsace.

Would the burden of bureaucracy in all these activities remain subdued and allow the accommodation of Gilbert's dream and also some space for independent thought, a loving family and a good game of table tennis? I hoped so, and I was tempted to answer yes, which meant Gilbert would reach the happiness he deserved. But then, I was only an *ami d'antan* ("old friend") and I had no access to the only unknown of the equation: the size of Gilbert's Dream.

---

31. Few months after the crisis described, Puget reached his retirement age. Before leaving Stockholm he called Gilbert and apologized for his accusations of Communism and (almost) treason: "*Voila', mon petit Karpman, vous avez les excuses du General* !"

Fig. 19 – Student revolution at La Sorbonne. May, 1968.

6

The Anti-DNA Story

DNA is the *Code of Life* and genes are made of DNA. This may sound like common knowledge today, but in the early 1950s it was a mere assumption, when only a small tip of the iceberg was known. Despite, or because of this, any new related finding would become the focus of attention and curiosity.

One of the things that led to further knowledge of the DNA was medicine's effort to find a cure for Leukemia.

In Milan, third-year medical students could apply to become "interns" in a University department. In 1953, I file my application to the Medical Clinic. The applicants exceed the openings: when I am accepted I feel like I have won the lottery and report to the laboratory of Hematology, one of the few labs offering hands on experience for medical students. Head of the Lab is Elio Polli, one of the senior Assistant Professors.

A pleasant but strong personality, years later he would be called upon to succeed the legendary Professor Luigi Villa as director of the clinic and number one internist in town. However, what makes Polli forcefully influence the first steps of my career is his conviction that the solution to the Leukemia riddle lies in the DNA and the hopeless patients that crowd his infirmary can only be saved if he succeeds in determining *what* is the difference between normal and leukemic DNA, and what went wrong in the latter.

Each intern student has four beds to tend to in the morning, but immediately after the "rounds" (a procession-and-discussion of cases by the professor and his assistants), I run to work in the lab. Our assignment is to purify DNA cells of different types of leukemia firstly from patients of the clinic and secondly from any source we can lay our hands on (calf thymus from the slaughterhouse, leukocytes from the blood bank) and confront the two.

I work singlehandedly and my equipment is state of the art, directly imported from the US, following the specifications of Prof. Erwin Chargaff, to whom Polli has paid regular visits in Chicago.

The tools are the following:

- 4 *Beckman Centrifuges* looking like flying saucers on tripods, which have to be run in the cold room;
- a *Waring Blender*, considered essential and the most difficult piece to get because nobody in Italy knows what it is (even if, actually, it is the equivalent of a common kitchenware appliance);
- a custom-made *Condensation Chamber*, the size of a large vase, with a rotating arm to collect threads of pure white DNA materializing in absolute ethanol at a temperature of -10° Celsius.

As soon as a preparation is made, I deliver it to a team of young doctors in the next room; they seem very competent and a bit status-conscious. Several of them (DiMayorca, Semenza, Basilico) will become internationally known basic scientists[32], but their analysis is crude: they measure the viscosity (glueyness) of DNA solutions at different pH (acidity levels). The probability of discovering meaningful differences by this method is low, and reflects what is known at this time (the age of the *Tetranucleotide Model*[33]): hopelessly little. Polli realizes it; he suffers and searches for new, more discriminating ways to fulfill his mission.

In the spring of 1955, a two-month studentship (*bourse*) issued by the French government is received by the University of Milan. The dean, considering the award too tiny for a regular contest, sends it to Prof. Villa, who in turn asks his assistants to find the intern student who has worked the hardest. The Senior Assistant, Dr. Breda, calls me into his office:

"You are lucky, the sixth-year interns are busy with their thesis, so the *bourse* goes to you. The dean's office wants to know which institution you want to be sent to".

That week Polli is traveling to the States, so I reply:

---

32. *Basic Science* is research aimed at understanding fundamental problems; *Applied Science* is the application of basic scientific knowledge to solve practical problems.
33. DNA, *Deoxyribonucleic Acid*, is present in the nucleus of all cells. Ervin Chargaff's model repeats the chemists' finding of the forties: DNA contains four *Nucleotides* (nitrogenous bases), two of them are *Purines* (Adenine and Guanine); two are *Pyrimidines* (Thymine and Cytosine). Purines interlock with Pyrimidines, making DNA a very stable molecule. The article *Molecular Structure of Nucleic Acids*, written by Watson and Crick and published on Letters to Nature in 1953, reveals the real function of DNA: to be, preserve, reproduce the Code of Life by using series of three bases signifying each of the twenty amino acids to compose proteins. The code is written on one helix, which also serves as template, a system that allows cell duplication and organism proliferation. It turns out that this Code is the same in all living beings on Earth.

"Polli is my boss, I cannot tell you until he comes back, probably next week."

Breda shakes his head:

"No, there is no more time. Tell me now, or we lose the *bourse* ...Villa will not be happy!"

After excruciating hours, I realize two months is not enough time to do any serious research projects and that perhaps I could explore some other specialty within the Faculty of Medicine in Paris, like: Cardiology... Pharmacology...perhaps Psychiatry? Why not visit all three?

Dr. Breda writes down these preferences, looking at me in sympathetic disapproval. A few days go by. Back from Chicago, Polli arrives like a fury:

"No nonsense! You are going to Pasteur! I just saw some slides showing that antibodies are able to distinguish between similar molecules. Therefore, *Antibodies* are the tools we need for the DNA project and you are going to learn all the techniques of DNA precipitation[34] from professor Grabar! His famous lab is at *Institut Pasteur!*

I see his light. Broken-hearted (my girl has just left me), I am more than ready to go. Dr. Polli asks his colleague in the Biology Department, Prof. Ranzi, who has once invited Grabar to give a seminar, to write him a letter on my behalf. It takes several trips to Ranzi's institute to get my precious letter: it is never ready. "I am busy with the exams this week: come next Wednesday!", he keeps

34. *Precipitation* is the method we use to determine if there is binding between antibodies and DNA. If the molecules bind, precipitation occurs and the molecular complex formed appears like a white cloud, separated from the rest of the aqueous solution, it becomes insoluble, and falls to the bottom of the test tube.

saying. Finally, time is running out and I decide to call him, something students would never do to professors. He says:

"I have written the letter. It's in the mail. Go, and best of luck!"; and, at the end of October, I am on the night train to the *Ville Lumiere*.

My arrival at Pasteur is unexpected. Pierre (Piotr Petrovic) Grabar has not received any letter.

"And, had I received it, I would have answered no, we have no openings!"

Then he looks at me and must see the misery in my eyes. After a pause, he says:

"However, I realize it is not your fault, and I know by experience how difficult it is to begin… (I will learn later that, as an officer of the White Russian Army, he had fled for his life in 1917, and had worked as a chemist in the mines of northern France before emerging at Pasteur) I will keep you, and you are assigned to the lab of Maxime Seligmann".[35]

---

35. Maxime Seligmann, recently graduated as MD, is writing his fourth-year Doctoral thesis with Pierre Grabar, at Institut Pasteur. He is using the methods of immune precipitation in gel that Grabar has perfected, and invented, to discover antigenic proteins inside human leucocytes: he can already distinguish more than twenty precipitation lines in the agar gel, meaning at least 20 different antigens. He believes that the distribution of these antigens in the population is genetically determined, and therefore hopes to find a correlation between these pictures and the compatibility of organ transplantation. At the moment, transplants can be exchanged only between identical twins; not even transplants between parent and offspring, or among siblings appear to be accepted. Jean Dausset, a former colleague of Maxime's, working at Hospital St. Louis, across the Seine, has similar hopes, but his method is cell agglutination by antibodies, a method that limits his field of analysis to leucocyte antigens that reside on the cell membrane. The problem is: the agglutination of leucocytes is too crude and involves only few antigens. "Jean will never make it despite his enthusiasm!" - explains Maxime, shaking his head, and resuming the analysis of his plates with spider web precipitations in the gels. I will meet the future Nobel laureate, Dausset, 10 years

Figure 20 - Instutut Pasteur

Young doctor Seligmann accepts me *obtorto collo* (unwillingly): what he objects to is not the foreign student, but being *obliged* by his boss: and he makes this point perfectly clear for an entire month.

On day two I am asked if I would please deliver a bottle of blood ("important for an experiment") to a collaborating hospital. I am

---

after the time of this discussion in interesting circumstances in Stockholm: (see in *Gilbert*, Chapter 5).

given scanty instructions: "*C'est facil, on prend le Metro a' Pasteur*". ("It's easy. Take the subway at the Pasteur station"). Three changes of subway are required and I have never seen the subway before.

When I reappear in mid-afternoon, I am looked at with funny faces (I must have been counted as lost). A few days later I am volunteered to test a glass beads device that takes separates *Fibrin*[36] from fresh blood and speeds up many lab tests. My blood is descending from the elbow vein and disappears in a dark red rubber tubing remaining invisible until it reaches the device, a blown glass cylinder full of beads, which is lying on the floor. When my blood begins expanding around the spheres like little red fingers, I, a healthy but low-pressured boy, faint.

Fig.21 - Franco is volunteered for blood donation

36. *Fibrin* is a filamentous protein that enables blood clotting. Fibrin is soluble in normal plasma (*Fibrinogen*) but when platelets sense that a blood vessel is hurt, they break and release their content (*Thrombin*) making fibrin insoluble to enable blood clotting. In the lab, clotted blood yields clean serum after centrifugation. Instead, in order to recover and study blood cells, the clotting process must be prevented: several methods are available

This is my victory. When I open my eyes, and see Seligmann's expression, I read it like a book. He knows he has trespassed his physician's boundaries. I was a youngster and a foreign guest: suddenly, through a simple needle, I have become his patient and his responsibility. He takes my pulse, which has recovered by now. I smile and he is off the hook. We will become the best of friends.

I do learn all the precipitation techniques [37], from the simple *ring test*[38], where the antigen solution is delicately deposited on top of the antibody in a thin glass tube, to Oudin's linear diffusion in gel (that affords counting the number of immune complexes formed), to the *Ouchterlony*[39] two-dimensional precipitation, revealing different

---

37. Most antigens are proteins, chains of Amino Acids, which always have "water-friendly" spots (*Hydrophilic*) and "water-repelling" spots (*Hydro-Phobic*). Only the former keep the molecule "wet", that is, soluble, and the size of the protein, of protein-complex bound together is important: the bigger it is, the more Hydrophilic spots are necessary to keep it "in solution". The effect of antibodies binding specifically to a given protein is to cover some spots, and if more antibodies can bind, to create "Complexes" of identical molecules: both effects concur and fatally the complex stops floating, a large number of complexes become a very large precipitate that goes to the bottom of the test tube. If instead of in liquid all this occurs in a gel, which is 99% water but has a maze of filaments keeping objects from moving freely and sinking to the bottom, the precipitation occurs anyway, but within the gel, forming a visible tridimensional image of the Ab/Ag reaction.

38. The *ring test* is the simplest laboratory procedure to determine whether a serum contains antibodies against a given antigen. The test is performed in a 2-mm wide, 4-cm tall test tube. The serum is introduced into the tube using a fine Pasteur pipette to fill the bottom half of the test tube. The antigen solution is then layered on top of the serum. At this point the interface between the two liquids appears as a clear disc (ring) when observed by the naked eye under a sharp side light. If the serum contains an antibody against the antigen, the clear disc turns into an opaque white line of precipitate immediately or within 15 minutes.

39. A novel technique at the time invented by the *Oerjan Ouchterlony*, which allows studying the precipitation occurring when antigens and antibodies bind in precise proportions. This test uses glass plates coated with a layer of agar gel in which 3 wells (0.5 cm in diameter) are cut at the corners of an imaginary equilateral triangle with 0.5-1.0 cm sides. One well is filled with antigen and two the

antibodies and different antigens and finally to Grabar's own *Immunoelectrophoresis*, where antibodies, diffusing through an agar gel, meet with antigens, which have been separated from each other by applying an electric field. It takes four months instead of two. My generous father thinks that the sum he has to remit by mail to keep me at the *Cite' Universitaire* for a longer time is a good investment. And four months at Pasteur are sufficient to catch an incurable virus: I will become an immunologist at any cost, even if I have to give up a physician's career.

The last week of my stay, Maxime Seligmann's wife, Francoise, asks her husband:

"Is it true that you have an Italian student in your lab and have never brought him for dinner a la *maison*? If so, you'd better correct this!"

The next day I am invited and Maxime drives me in his *Deux-Chevaux*. The car traverses parts of the *Sixieme Arrondisement* - I recognize the Jardins du Luxenbourg - and stops at the curb. I ask:

"Is this where you live?"

Maxime says: "Very near, opposite the park, but I stopped because this is the last moment we can "talk shop" together, and I have a couple of questions for you:

"Where is this "DNA" that you always mention?"

"Well, in the nucleus of all cells…"

"Why is it important?"

---

antibodies or, alternatively, 2 wells are filed with antigens and one well with antibody, and antigen(s) and antibod(ies) are allowed to diffuse in the gel all around the corresponding well. When the antibody meets the antigen at the edge of the diffusion ring, the two molecules bind to each other, forming a complex that precipitates in the form of an opaque arc. Precise and interesting conclusions can be obtained by analyzing the position of the precipitates formed.

"Dr. Polli is certain that leukemia is caused by some fault in the DNA..."

'Why do you want to make antibodies against DNA?

"To distinguish the "good" from the "bad" kinds using precipitation"...

"Are you sure that DNA will elicit antibodies?"

"No, I am not, but I hope it will..."

"How can one purify it?"

"I am the expert here...we follow Chargaff's procedures"

"What is the *Molecular Weight*?"

"Many millions..."

Lupus is not mentioned at this time, but we are both aware that it is a mine for *auto-antibodies* and that anti-nucleus antibodies[40] have been described in Lupus patients together with all sorts of auto-antibodies, as if all provisions for not damaging SELF were cancelled in these autoimmune patients. This is a lead even if still without a clue.

Back in Milan, I need an animal facility to host the rabbits I am going to inject with DNA. The Medicine Department is not the right place. Polli calls human geneticist Ruggero Ceppelini: he agrees to let me work afternoons at the *Istituto Sieroterapico Milanese*, a venerable private institution in the outer periphery of town.

I travel miles on my faithful 48cc 4-stroke Motom (a jewel for its time), inject DNA from various sources into eleven rabbits, with and without *Freund's Adjuvant*[41]. Months pass. My girl unexpectedly calls

---

40. Researchers must give a name to what they find. Anti-nucleus antibodies were so labeled because some Lupus patients' *sera*, incubated with normal cells caused a degeneration of their nucleus, similar to the one seen in Lupus cells.

41. *Freund's Adjuvant* is a concoction containing oils, proteins, and boiled bacterial shells. If any antigen is injected in emulsion with the adjuvant, the antibody response is much more abundant than if the antigen alone is injected. Charles Janeway labeled Freund's adjuvant "the dirty little trick of immunologists"

me. Her art studio is halfway between the clinic and the lab. My Motom begins making an extra stop. Some inconsistent shadows appear and disappear in the agar gels instead of the Immune precipitate that I expected. The conclusion slowly becomes clear: DNA is not *immunogenic*[42]: it is not capable of inducing antibodies against itself. This is anticlimactic. I have invested a lot in this project on which I planned to write my thesis. I continue my routines like a robot.

One morning a new patient appears by chance in one of my four beds. It is a severe case of *Lupus Erythematosus*. I collect his medical history. That afternoon I bring a sample of his serum to the lab. The same evening, I try a ring test: layering a DNA solution onto a series of dilutions of the lupus serum. The tubes are hand made on the fire, cut to be 5cm long and 2mm wide, and stuck vertically in a piece of clay.

The white precipitate at the interface looks like a slow-motion explosion in the side-illuminated test tubes. A real thrill; my first. Ceppellini, who has acted skeptical about the rabbit work, becomes instantly interested when he sees my Lupus precipitates. If there is a success in the laboratory, he wants to be part of it. He orders his faithful technician, Mario Turri, to perform a *Complement Fixation Test*", which turns out positive, the surest proof that an immune reaction of the body has taken place.[43]

---

because its use prevented for decades the understanding of the physiological controls that allow immune responses only against dangerous antigens and invaders.

42. *Immunogenic* is any molecule that, when injected into or contacted by a vertebrate organism, can elicit a specific immune response. While all immunogens are antigenic, not all Antigens are Immunogenic. Why? This was a puzzle in 1955, and will be fully solved ten years later.

43. Ceppellini is flabbergasted because he has remained skeptic about Polli's idea and Celada's endeavors: now, seeing the positive ring test he must accept the result; furthermore, as a brilliant geneticist, he imagines that anti-DNA antibodies

This is a step forward in immunology. The presence of anti-DNA antibodies is a sure diagnosis of Lupus, while its absence in normal non-Lupus individuals means that the immune system goes a long way to actively block potential dangers, as well as useless responses. Instead the Lupus disease will be considered reactive against all possible antigens, a multi-faceted self-immune condition. The presence of DNA-binding antibodies (i.e., anti-DNA antibodies) in the serum of Lupus patients was reported by three independent groups in 1957, from Milan, New York and Paris. Science does not operate in a vacuum, and, ironically, the thread that linked the three discoveries was the youngest and least experienced of all the researchers: still a medical student, I happened to play this role unknowingly.

Fig. 22 – a) Professor Elio Polli, Head of the Hematology Lab of the Medical Clinic of the University of Milan. b) Porfessor Piotr Petrovic Grabar, Chief of the Immunochemistry Lab, Institut Pasteur, Paris. c) Dr. Maxime Seligmann, student of Professor Grabar

---

found in his own backyard will have important applications, which he will be able to guide and control. The complement fixation test that he suggests, and asks his technician to perform is a classic test and will show that the precipitation in the ring is caused by antibodies.

Fig,23 - The labs of the top floor at the original Institut Pasteur, built in 1868, were awkward because of the mansard roof hovering over the benches; Also, the bench assigned to me was occupied by a large apparatus, with whom I cohabited for four months: a shiny upright cylinder closed by two plates, top and bottom, secured by an excessive number of bolts and nuts. It was a classic filtering devise, which would act under pressure and using any sort of filtering material. Impressed on the steel was the label: *Seitz Werke DE.* At mid-century the Pasteurian were preparing all materials needed for the experiments: glass pipettes were blown on Bunsen flames, ad all sorts of solutions were produced, and had to undergo *ultrafiltration* to eliminate any impurity and microorganisms. Thinking of the hectoliters he must have purified in his working life, my respect for my friend Seitz grew. When my *Burse* came to an end and the people in the Grabar department threw a little party for me, my impulse was to promote Seitz to the rank of Professor ad invite him to my goodbye event. For the occasion I dressed him up with a tie and lab coat, giving him eyes, spectacles and a mustache. The following illustration above is faithful to the photo on my Kodak.

Fig.24 - Brothers, or first cousins are the resource if you need a kidney transplant, since they may produce a perfect match, with 50% or 25% probability…but if you seek something even more precious, like true friendship and companionship, then it is the third cousins, with their numbers that will show the best selection! Riccardino Besio, two years my elder, offspring of my mother's best cousin and a naval ship manager at the harbor, just continued our family saga. During the years, we have been meeting in Genoa, by the seaside, in Milano, in Los Angeles and in Paris. Riccardino was in Paris for business and his stay overlapped for two days with my adventure at Pasteur: enough for a needed soothing effect. Once we literally "ran" into each other at the Champs-Élysées in the midst of the celebration of *La Toussaint*. It was October 1, 1955.

7

October '56: Looking Forward

I stand in front of eleven professors of the Medical Faculty of Milano, crammed behind a long desk, wearing awkward black capes lined in red. They are surprised that I present original work (injecting 17 rabbits, collecting dozens of Lupus patients' samples, testing them with techniques imported from the *Institut Pasteur* in Paris) and that I even obtained a publishable result. No wonder: most of my fellow students limit the thesis effort to a review of the literature, since there is no time to squeeze research into the tightly packed curriculum of the sixth year of Medicine. As a rule, Western nations, with healthy pragmatism, do not require a candidate medic to discuss a thesis in order to be allowed to practice medicine. Idealist Italy does, and the 1924 university reform introduced it, to show that Italian medical doctors are better than their French, German and English counterparts. It was a typical fascist idea, combining wishful thinking and make-believe, and then failing to allot the means. Mussolini was then in his second year in power and did not say how a mediocre thesis could produce a better physician. It was left to us youngsters especially those who chose world-class research for a career, to taste the bitter truth, since no change in academic thinking had emerged ten years after the demise of fascism, and it wouldn't happen after fifty years either. The allotted time slot for my thesis presentation (about five minutes) is already over. The president of the commission asks whether there are questions. Professor Margaria[44] raises his hand:

"Nice work. I move that your thesis be accepted with honors. But what are you going to do with your anti-DNA antibody?"

I do not know the answer, and I am pissed. This question increases my urge to continue the work, now that my paper has been accepted by the *Society of Experimental Biology and Medicine*. But there is a big boulder on the road, ready to shatter my plans: the military service. It is compulsory, and will rob me of eighteen months of research life. How can I get off the hook? Perhaps a health condition could make me non eligible! But I am not the only child of a widowed mother, like Raineri; I have no tiny hole behind my ear, remnant of embryonic evolution, like Pisani.[45] I have no severe

---

44. Margaria taught biochemistry in second year and physiology in third, my favorite courses. He was smart, interesting and often sarcastic. When Raineri and I could not understand a page of the physiology textbook "Margaria & De Caro", we decided to go and see him. Professor Margaria opened the book and asked: "Which page did you not understand?", then: "Oh, I see, this part was written by De Caro!" He read a few lines, then closed the book and uttered: "I don't understand this either! Just forget about it and study the rest! See you at the exam!"

45. The Pisani story had a special twist. His dad was a professor of Urology and his dream was to see his son succeed him on the same chair (nepotism was and is reigning in Italy), and feared any possible doubt about his physical fitness be raised and disturb the boy's future sailing in Academia. Thus, he wrote a pleading letter to the Military commission, and contacted the examiners one by one, urging them to enlist the candidate. The Military however, accustomed to pressures and even bribes in the opposite direction, became suspicious, and followed the law, which made the young Pisani unfit for the Army as "carrier of a genetic malformation". Pisani's abnormality, was a gram of embryonic tissue behind his left ear, a remnant of a normal process called *Recapitulation of Phylogenesis in Ontogenesis*: the appearance in the fetus of traits reminiscent of earlier species in the history of life

ailment...wait a moment: what about my knee? After the somersault last year, and the crazy immobilization in plaster for forty days it has given me only trouble. Could this be a chance for its redemption? Could it save me from the 540 days I risk spending in the Army? The next morning, I travel by tramway to the Military Hospital of Baggio, in the periphery of Milano, where the medical officers to be must pass their physical. Long wait, then a door opens and my name is shouted. I enter and look around. Leaning on the examination bed is Paolo Mombelloni, dressed as a Lieutenant under the white coat. A real gift for me. We belonged to a small group of student/friends, who followed the same courses from years 1 to 6. Then, when I took time off to do research, Paolo went straight to the Military school and is now in the last months of service.

"Paolo! What a surprise to find you here!"

"It is a surprise for me as well; they just deployed me here for the tests! And now I will be your examining officer! How are you? I heard from the group that you have a discovery!"

Paolo puts on the stethoscope and gives me a routine check-up: blood pressure, heart, lungs, and pupil reflex to light...

"You are in splendid form, as I expected!"

"Yes, but how about my right knee? Don't you remember how long I kept hopping around with a stick?"

He looks at the knee, tries to bend it. It hurts at 80 degrees, and cannot be completely straightened.

"What happened to it? It was tennis, if I remember correctly"

---

on earth (gills like a fish, tale like in amphibian), which as a rule, disappears well before birth. There is no harm involved in this display. It is mother Nature's reminder that life on earth is one and we are part of evolution. A joke that only myopic Creationists have problems accepting with a smile. Pisani became, as everybody expected, professor of Urology at the University of Milano.

Fig. 25 - Lt. Paolo Mombelloni evaluates Franco's fitness for the draft

"No, it was a somersault *after* tennis, to celebrate the Men's Double victory in the Monte Rosa Tournament."

Paolo erupts: "*Allora te la sei cercata! Ma chi te l'ha fatto fare!?*" ("Then you really asked for it! What the hell were you thinking?")

"I must admit, it was a show for the girls of the band. They loved jumps." Paolo becomes serious:

"Franco, it is really up to you. For once, my conscience and my obligation to assist a brother doctor under Hippocrates' oath are satisfied with either decision. Your right knee is hurt but improves. I can either write that it is ailing and therefore you are not fit, or choose to ignore it, being convinced that it will heal spontaneously in a reasonable time, and declare you fit."

A devout Catholic, Paolo savors this spark of infallibility.

"Well, thanks a lot. Then I will choose to be unfit and continue my project!"

I am ready to shake hands.

"Not so fast, there is a catch: if you are rejected as officer, you are not yet off the hook: you will be tested as a private, and there you may be found fit."

I had not thought of that. With fury I say:

"But medicine is a Science! If I am sick as an officer I am sick as a private, don't you agree? He smiles and answers calmly:

"You are wrong here: first of all, the requirements are higher for Officers than for Privates. Secondly, the examiners in that commission are likely to be hard boiled and suspicious career Officers and, third, you will be called in a couple of months and by then your symptoms may have faded. In conclusion: you are still free to choose, as I told you, but it is a risk that I, in your shoes, would not take. Imagine living one year and a half as a Private and Private is only one notch above shit. Instead, as a Lieutenant you even get paid, and you have status, since medical opinion is "holy": not even a General can touch it!"

I hate Paolo's suave tone, but I realize he is right in his reasoning and I am trapped. I enroll. That even makes me technically a volunteer.[46]

This year is full of deeds and decisions. I am on my way to the *Istituto Sieroterapico*, where I still have my rabbits. I will abandon them when I go to the *School of Military Medicine* in Florence, but first I want to talk to Ceppellini[47] He has changed his attitude lately and is extremely friendly to me. I am uneasy since the change coincided with Polli offering him to become the first author of our paper if he would translate it into English...Today he is ebullient. When he sees me at the door, he shouts:

"Come in, Celada! Or should I say *Doctor* Celada? Congratulations!"

He wears his usual brownish tweed jacket and helps his reasoning with large gestures:

"First, you should not worry about the loss of time, what's important for you is to have a good plan ready when you are released. And you are lucky: when I was drafted in 1940 they sent me straight to Libya, the boat was sunk by a torpedo and the British

---

46. In peaceful Italy, the18-month draft was compulsory for all at 20, while a delay was allowed for university students. Medical graduates had the option to volunteer as military doctors and, before practicing, they had to learn "military medicine" (as if such a thing existed), taking a four-month course given in Florence.

47. Ruggero Ceppellini is an Italian Immunogeneticist, Professor of Human Genetics at the University of Torino. As a person who had a big influence (perhaps too big) on the early stages of my career and family life, Ceppellini is a recurring character in this book and additional information about him can be found in the text of several chapters (*The Anti-DNA Story*, *October 1956*, *Blockhead*). I think that, as a book character, the man is an asset; as a mentor, boss, or co-author: no comment.

took me as prisoner: I came back after six years! Second, there are great developments. I am practically sure that when you return we will not be in this crumbling institute any more. I will become a Professor and part of the merit is of our Anti-DNA![48]".

This our rubs me the wrong way. He sees my puzzled expression, and rushes:

"I tell you why!", as he begins to open the new issue of *Nature* on his desk. He finds the page and hands it to me:

"Look for yourself!"

The magazine is dated September 8, 1956. And the article is by the trio Billingham, Brent and Medawar, that have been stirring Immunology during the last five years. I begin to read the title: *The stimulus in Transplantation Immunity.*

"It seems rather definitive!"

But Ceppellini is impatient, and cannot wait for me to read the whole article. He repossesses the journal and shows me a spot:

"Here, look what they say: *...the antigens are certainly nuclear substances*, and then, in the conclusions...*the antigens responsible for Skin Transplantation Immunity are Deoxyribonucleoproteins*...you see? Do you realize what this means?"

I venture:

"Transplantation antigens define the individual, it makes sense that DNA..."

I am interrupted by a bull's charge:

"It means that there will be a worldwide race to crack organ transplantation, but only we have THE tool: anti-DNA antibodies!" and concedes: "*your* anti-DNA".

Well, what can I say.... this man must be a genius. I am overwhelmed. This is not the moment to point out that our Lupus

48. See *The Anti-DNA Story*, Chapter 6.

anti-DNA binds many kinds of DNA, in fact all the kinds I tested. Ceppellini savors the effect of his tirade, lights a cigarette, inhales, exhales, then continues:

"There is another coincidence. Last week I met Alexander Hollaender. He gave a talk in Pavia, and Buzzati introduced us."

I know the name of geneticist Adriano Buzzati Traverso, but I never heard of Hollaender. Ceppellini explains that he is the founder and head of a division of the *Oak Ridge National Laboratory* in Tennessee, the place where they made the atomic bomb.

"Since the effects of radiations can be cured only by Bone Marrow transplantation that institute has become THE world center for transplantation research."

I wonder where this talk is heading. I am going to learn it soon: …"and Hollaender has asked me to come to Oak Ridge and set up the lab of immunogenetics!... Naturally you are on board".

"You mean, I am also invited?"

"No, I am invited, and I'll take you with me because you have proved to be perfect for the project. You will have to apply for a NIH Post-doctoral Fellowship. With the Medawar background and the anti-DNA, success is guaranteed. And, of course, I will help you write the project. What do you say?"

I think: What a genius, but what a charmer …! I say:

"Of course I like the idea! But what a nuisance this Milit…"

He cuts me short:

"I told you, this is a false problem: it takes time to write the application. There are fixed deadlines for sending and Italian applications must pass first by a commission in Rome, and then there is the processing time. It takes about a year to get it: a good use of a soldier's life".

Back on my Motom (see Chapter 6 , *The Anti DNA Story*) on the Milan beltway, rushing to see my fiancée, Antonella, and give her the great news, my thoughts on Ceppelini's "show" are mixed. A great vision! A great seducer! Any hidden dangers? Clearly this project serves his interest...But, I must admit, also my own: for the first time I may have work and salary, and...enthusiasm is building up.

The eighteen months of military service felt like I have been crushed by a ten-ton cement block out of a Monty Python cartoon. In mid-august three hundred doctors enter a former convent on a hill near Florence as trainees to become officers and come out six months later, properly dressed and labelled as medical second lieutenants, ready to be "shipped" to hospitals, barracks and maneuvers. I convince myself that I can use a change of rhythm after the pressure of the last two years, and that I may even learn something useful after all, while keeping an independent mind. The latter proposition turned out to be the most difficult to achieve, but I did learn a few things.

LESSON 1: HOW TO PROTECT YOUR NAME.

At the onset, my name has an incredible value for the army Officers, who command the company of doctors: the Captain in command and the three lieutenants, one plus a Sergeant, for each platoon. At the beginning, the officers have no grasp on us, and no power onto an anonymous mass of trainees: they need a handle, a name, not only to a face, they can associate to some breaking of rules, or any funny, or stupid happening. Asking for names at random does not work, but conquering one or two names per platoon suffices. Once the name is caught, it becomes enshrined in the officers' memory, and is ready to be used as soon and as often as needed. I am unaware that my name is among the first ones to be surrendered: the occasion is a cascade of paraphernalia falling out of my locker at the very first inspection: books, letters, racquet and tennis balls. The next day platoon three is marching in Piazzale Michelangelo, one of the marvels of Florence: all towers, cupolas, gardens, bridges across the Arno are spread in front of us. Lieutenant Sasso governs the steps by uttering guttural inhaling-exhaling sounds: "*Op-due, Op-due*!"[49], but prefers to lean against the parapet and rely on his resonant voice, instead of marching with us.

"*Op-due', Op-due'*!"

Fig.26 – Soldiers' drill at the Military Medical Academy Costa San Giorgio. Find the odd man.

49. "*Op-due*" is the sharper condensed version of the original marching command "*Un-due*" ("one-two"). It is more efficient in open areas like Piazzale Michelangelo and the lieutenant can control his soldiers 300 hundred feet away.

The platoon is now two hundred meters away. Sasso sees some irregularity in the up and down oscillation of the heads. It is impossible for him, at this distance, to see who causes it. But he shouts my name:

"Celada! In step!"

Whether or not it is me (and it is me this time) that disturbs the dynamic, geometric rhythm of the marching unit, is irrelevant: the entire platoon is electrified and under Sasso's grasp. And everybody feels innocent and associates my name with the faulty march.

In the afternoon in the large arch-lined court, before the *libera uscita* ("off duty"), it is the captain, dutifully informed, that calls my name to remind the entire company to follow the rules of attire. After two days my haphazard crossing of another court is greeted by a moron member of the platoon, who spells out my name and ten other sheepish morons who repeat it, shout and laugh and laugh. What a lesson I learn! Unfortunately, it is of no use for me: naively, I have fallen into the trap. There is no way to repair, or undo it.

LESSON 2: TO STITCH AN OPEN WOUND.

I have the perfect occasion to learn this skill in Verona. One night I am the attending medical officer. I am alerted by the sound of an incoming ambulance. A private is carried in ambulatory on a stretcher: he had an accident in a truck. The wound is superficial, but lurid. The skin of the right half of his forehead has been cut by a broken glass just below the hairline, detached from the bone and flapped down like a pita bread cut horizontally by a knife, covering the right eye. What I see is a double surface of bloody exposed tissue. We clean the wound and wash it with antiseptic. When the skin is repositioned into place the soldier's face becomes more human and reveals a clean curvaceous cut from just over the nose to the right temple. It must be sutured. Resident Sargent Major Terranova hands me the needle with attached thread. I ask, thereby loosing half of my standing:

"How many stitches will it take?"

"At least thirty" is the matter-of fact answer.

It is easy to stick in the curved needle, but difficult to penetrate the detached flap from the inside out to recover the thread. I make two stitches. I feel little drops of sweat on my forehead. Somebody must have forgotten to teach us this technique in medical school. The patients' eyes express only trust in me. I can't stand it anymore. I ask the Sargent Major to please take over. His title is male nurse, and despite the enormous size of his body and his hands, he does an excellent job for the relief of two: the soldier and the doctor. I have lost the second half of my standing and I do not learn this time.

LESSON 3: TO GET AN ASSIGNMENT NEAR HOME.

After the course in Florence the medical lieutenants will be assigned to one of the five military zones of Italy: North-West, North-East, Center, South, the Islands. Everybody hopes to be assigned near home but the criteria for the decision is a secret jealously kept by the authorities. Three factors are believed to be important:

1 - High marks in the course *Attitudini Militari* ("Military Skills")[50];

50. Military skills are what the three hundred graduates, spending four months in the school of Florence, are supposed to learn from scratch. While Military Medicine is what they have studied for six years in the universities, as applied, adapted and often distorted for military use. At the end the candidates will be judged on both their scanty performances (marching, shooting, cleaning the rifle) and on their knowledge of the army structure, corps and hierarchy and on the theory and the effects of the last decade's novelty: the atomic bomb.

2 - A *letter of recommendation.* The power of the latter depends upon the rank of the writer, who, in descending order, could be a:

a) *Cardinal* (the highest ranking is Cardinal Montini of Milano, future Pope);
b) *General* (not if he is retired);
c) *Cabinet* member;
d) *Senator* belonging to a majority party;
e) *Star Surgeon*;
f) *University Professor*

3 - A *direct communication* of the "recommender' (in person or via telephone) with the person responsible for the assignments of the soldiers.

The extremes to which my colleagues are prepared to go to secure one of these "factors" are unbelievable. Some give up the evening leave to study military attitudes (sic). A few study by night: since it is prohibited to use portable lamps in the bedrooms, they construct a tent with their blankets and read by candle light. The results are: red eyes in the morning and some sheet with charred marks. The tension increases by the week. Signori is the victim of a cruel practical joke. He storms into the captain's office agitating a sheet with a list of names:

"Signor Capitano! I protest against injustice! This rank of top thirty in military attitudes was released before the final test! And my name is not even there!"

The captain examines the document. It is our barber's list of haircuts for the day.

I have no recommendation letters and my record in marching and rifle cleaning and handling is pitiful. I am out of the race and I feel free to ridicule my undignified colleagues.

I will end up in the North-East region, not too far from Milano: the *Military Hospital of Verona*, built by the Austrians a century ago, featuring a majestic winding staircase, with extremely shallow steps, built to allow horses to pull carts and beds up to the third floor in the 1800's.

LESSON 4: HOW TO *DISSECT* MEDICINE.

The first separation is between future *Internists* and future *Surgeons*. Every single newly graduated doctor carries the "I" or the "S" imprint (researchers belong to the first group)[51]. The second separation is much more specific but equally net: *Military Doctor* (career Officer) and *Doctor*. Medical practice relies on a straightforward relation between patient and doctor: the patient seeks cure and trusts the doctor, the doctor depends on the patient for information, on which to program the cure, and trusts the information to be true. Military medical practice is an exception to this correspondence of agendas, and both career military doctor and newly enrolled medical lieutenant know or will soon realize it. Mutual trust cannot exist because now and again the soldier

51. Surgeons' brains are wired differently from those of internists, and the vocation is revealed in the first years of medicine, or even earlier. When they convene in Florence, the decisions have already been made and it is funny to realize that a few minutes of conversation are sufficient to guess the future career of any just met candidate. Medicine needs both varieties: those capable of making quick decisions and act without regret and those who are open to doubt and go to the "what ifs" suspecting that medicine and nature itself are, at the end, unconquerable. The future surgeons, like the barbers, with whom they have been associated in the middle ages, have fulfilled their duties and have the great advantage of physical solidity, a sure hand and resistance to fatigue.

reporting sick wants not to be cured but to obtain something else from the doctor (dismissal, rest, whatever). On the other hand, the doctor must take into account that the answers to his questions may be dishonest and unreliable. In response to this predicament two distinct attitudes emerge:

1) Major Carrubbo, head of Internal Medicine of the hospital of Verona becomes a policeman: for him all soldier patients are potential "fakers" and his principal mission is to seek them out.

2) Lieutenant Celada, his assistant, tries to distinguish the honest from the fake, but he admits to himself that his classic and academic education is no great help in this circumstance: therefore, when in doubt, he chooses to cure. He knows he might be fooled, but consider this to be a lesser evil than denying treatment to a private in need on the basis of suspicion.

The two medical men are never in agreement. They stage lively theatrical performances in the middle of the infirmary, a gallery shaped room lined by forty beds, each hosting an interested spectator. Remembering the encouragement of Paolo Mombelloni, I am not intimidated by the major and express diagnoses and propose therapies. But Carrubbo has a secret weapon: when he feels he is losing the argument, he commands:

"*Lieutenant! A-ttention!*"

From the beds, eighty eyes focus on the Lieutenant, whose voice freezes in his throat, whose arms automatically straighten, whose shoes clap, all as part of an acquired reflex.

Fig.27 - Lieutenant Celada is called to attention by his Major

I am a defeated, muted Second Lieutenant. But, wait and see! I am rather stubborn when I feel pushed. I have the support of my patients, and the trust of the nurses. It is now 11 am. In a few hours

I will be in charge of the afternoon rounds (Majors work only in the morning) and I will prescribe therapies according to my free conscience and my free will. The patients will get their full therapy. *Ave, Aesculapie!* (tr. Hail, Aesculapius!)[52]

LESSON 5: HOW TO SHOOT A CANNON.

Learning this does not take long. The artillery group I am assigned to for the period of summer maneuvers is equipped with 155mm cannons, which are now deployed in batteries of three placed at large distances between them in order to limit the probability of being wiped out by enemy fire. The soldiers want to see if their new doctor has the guts to fire one of the programmed shots. An officer hands me the string. I pull it at the signal. The shot goes off. The cannon jumps and retracts. My ears are overwhelmed, and for ten minutes I am deaf. At the end of the shooting session the artillery Colonel reviews the results of their test. Apparently, all the shots fell within a *fazzoletto* (literally: "handkerchief", i.e. a small area). The Colonel is about to congratulate his staff, when a Jeep arrives from the valley, carrying a civilian who wants to talk to the Colonel. The passenger is a local farmer. He takes off his hat, hesitates, then utters:

"My cow is dead, the cannons killed her!"

I begin to feel guilty: may it have been my shot? The Colonel seems upset:

"Mister, it is your own fault! We spent all yesterday and this morning warning all farmers of the region. Why didn't you pay attention?"

---

52. *Aesculapius* is an iconic Greek hero and god and represents the healing virtues of medical arts.

The farmer says:

"I did too! I rounded up all five cows, but she ran away, la Rossa ran away. You know, she always had a mind of her own."

Fig.28 - The farmer's cow, La Rossa, is hit by a cannon

"Ok", says the Colonel, now calmer but eager to avoid a debate on cow psychology. "I understand, do not worry. I'll sign the order for reimbursement immediately".

I am surprised by his sudden friendliness and generosity. The farmer returns to the jeep, appeased because the army pays well. When the jeep is out of sight, an applause erupts. All officers and privates are clapping. The Colonel wears a wide smile and caresses his mustache. A bottle of wine is opened for the occasion. I learn the following: military power is defined by its capacity to kill and destroy. However menacing, the fire-spitting artillery doubts its own power when, in peacetime, it is not allowed to hurt anybody. This accidental killing is a boost to its morale. The price of *La Rossa* will remain in the corps' records. The Colonel knows that it is money well spent to reassure officers and soldiers. The better I learn this, the more I develop my difference.

Fig.29 - The Colonel is ecstatic

LESSON 6: HOW TO CURE A COMA.

The city of Vicenza lies in the valley at the foot of the first terrace of the Monte Grappa mountains, called the highlands of Asiago, at 1000m of elevation over the Po Valley. Today, a two-hundred men infantry exits its barracks and crosses the town of Vicenza, marching towards the mountain in rows of four, fully armed and equipped, to the astonishment of the local population. Captain Goffredo Del Dongo, 42, tall, handsome, athletic and a nobleman, leads the march. He has planned it: in absence of wars he needs to collect some athletic glory to advance his career. The day is sunny, too hot; marching in these conditions is a momentous effort. With ten minutes rest every hour, it takes six hours to reach Asiago where the battalion has its summer quarters for the maneuvers.

Fig.30 - Soldiers' march from Vicenza to the Asiago

The marchers are greeted as heroes. There is a large table in the court of an occupied farm. The Colonel makes an epic speech reminding the many battles fought in a string of wars in this very region, and the deeds of the Captain's ancestors. It is a John Ford movie. Finally, everybody sits down for dinner. The first spoonful of soup reaches Del Dongo's lips. His eyes open wide. He falls from the chair in violent convulsions, unconscious.

Fig.31 - Captain Del Dongo falls into a coma

The Colonel retracts the spoon from his own mouth. Nobody touches the soup which is the presumed culprit. Since the battalion's doctor is on leave, a jeep is dispatched to fetch the doctor of any other unit in the maneuvers. The drivers must have had a course in acrobatics, judging from their evolutions in the tiny streets of Asiago.

Less than ten minutes pass, and they return with their pray, the aloof-looking Lieutenant doc of the artillery, yours truly. They take me to Del Dongo, who is now on a camping bed, where four of his men try to keep him put; strange laments from his throat, he breathes loud and irregularly, his heart thumps. I have never seen a patient in this acute condition and my mind is blank. To gain time I walk back to the dinner table. The Colonel and all officers are still there, in shock. All eyes are on me. I ask for a spoon, immerge it in the soup and put it into my mouth before anybody can attempt to stop me. It is still warm, delicious. From the taste and the timing of the accident I declare the soup (and the cook) innocent of all wrongdoing. I return to my patient and try to speak in his ear:

"Captain! Have you had this before? Answer me!!"

I pray he can hear. Seconds and minutes pass; it feels like an eternity. His crisis has ups and downs. In a moment of calm, he is able to whisper a broken word:

"*In-su-lin..*", before the convulsions resume.

It is enough! I take it as some gift from heaven: the man is a diabetic, he is under the effect of insulin. Obviously, he did not imagine that the march's physical exertion would cause sugar level to fall abruptly.[53]

"Sargent, I need 5% injectable glucose solution and a large syringe or an infusion device."

The Sargent runs, but comes back desolate:

---

53. There are several medical conditions that may induce coma. Diabetes caused by low or absent insulin production in the pancreas is the occasion for both high-sugar coma and low-sugar coma. The latter can happen for a wrong dose of insulin injected or in case of extraordinary muscular exertion, which multiplies the effect of normal insulin dose. This was the case of Captain Del Dongo. A rapid and correct diagnosis of coma is essential in order to apply the right remedy, which can be excruciatingly arduous.

"Sorry, Doctor, we are ill equipped: no glucose at all…"

"This is incredible! Call the cook!"

All privates around stare at me (and think: "another officer is going mad!"), but the cook, a corporal, arrives promptly and I tell him:

"Look, I just saved your ass, but now you must work for me: take 1 Kg of sugar, dissolve it in 3 liters of water. It must be heated and stirred to go fast."

Three men are involved, the cook, relieved, energizes his team.

"Then, we force the Captain to drink as much as he can!"

"Three liters?"

"Don't worry! Most of it will be spilled on the floor before he calms down."

The procedure begins. It takes a while to see any effect. Then, finally, Del Dongo becomes progressively calmer, he understands better, I can explain to him what has happened. The Colonel comes into the sugar splattered room and cannot stop thanking me for saving his Captain and much more.

The next afternoon I go back to the infantry camp. Their regular doctor has come back. He is a first Lieutenant because he chose to be a career military doctor. He has heard that I did well but he does not believe the Captain is diabetic, after all.

I shout:

"What! It was a textbook case! And sugar saved him from *Hypoglycemic Coma*!"

But I do not insist: military doctors are impermeable.

"Can I see Del Dongo?"

"He is not here. He was sent with an urgent mission back to Vicenza. This morning, by car."

"Can I see the Colonel?"

"Sorry, he is out in the field, directing the maneuvers."

"Then, farewell, my dear colleague."

"I am not your colleague: my grade is *First Lieutenant!* I am your superior!"

"I see, and I suppose that is why you know which ailments career officers may and may not have. And which, like diabetes, must never be mentioned or diagnosed! I am glad you are not my colleague: the little I know, after all, is just medicine."

Well, this time I really learned a whole lot.

In March 1959, the military duty ends. It felt like coming out of a long flu that left me breathless for some weeks. At first, it seemed impossible to recover completely, but then, back in Milano, the entire eighteen months of service seamed to fade away fast and disappear from my life. At home I find two letters: a yellow, official envelope from NIH and a note from Professor Ceppellini urging me to come and see him to discuss important issues. NIH, the number one science agency of the world, has awarded me a two-year fellowship and instantly transforms mere hopes into solid rock. The rock is located in the hills of Tennessee, it is far away, but not any less real. By focusing, I see all the components that make the rock tangible, heavy and precious: there is my first job, a laboratory shaped like a cube, covered with stacks and no windows, there is a marriage now feasible that transforms Antonella and me from a long standing promise into a family soon to grow. I run to see Ceppellini: the large envelop makes me feel prepared for any conversation even if problems should arise. This time Ruggero Ceppellini is not happy, despite his conquered chair in Academia: *Nature* has published a second letter to the editor of the British trio: Billingham, Brent and Medawar. It is much shorter than the first, and consists in an apology to the journal and all its readers: the idea that transplantation antigens are DNA[54] was wrong, and the authors hereby retract it.

There is a silence.

I am surprised by the news, but not shocked. Actually, I feel unexpectedly relieved: there must have been a subconscious doubt in my mind all along. Ceppellini is furious, shouting against this "devious way of making science", that has broken his clever scheme to use anti-DNA in this research.

"At this point it is clearly impossible for me to go to Oak Ridge, the Brits have undermined my project.... And you? Are you also renouncing? In that case I will make arrangements to get you a lab here very soon..."

I stare at him: is he crazy? Renounce my yellow envelope, and all its contents? Never! And, strangely enough, I do not feel any panic for being left alone in the American adventure. I tell him so in smoother words. Ceppellini must have expected my decision, and accepts it, promising that there will be a place for me in Torino after the fellowship:

"By that time, you will have acquired new skills and techniques, which will be an asset for my new institute!".

This last sentence contains a warning about the possessive nature of the man. Do I fail to see it? Or do I brush it away, rapt in my enthusiasm?

---

54. See *The Anti-DNA*, Chapter 6.

8

Piccapietra

Genoa, Winter 1978. Genoa is considered mild-weathered, but in this season, it is swiped by nasty wind gusts, blowing round corners, columns and passages. It is late afternoon: with the collar of my shearling tucked around the ears, hands buried in the pockets, I wander in *Piccapietra* looking for the sign of the restaurant where I am invited for an early dinner. This historic neighborhood, nested within the walled city, has lost most of its charm after the recent speculative renovation. The place is empty, modern and does not agree with the rest of the town. It is not surprising that I don't find a soul to ask directions. Finally, I see the words *Il Picchio* on a glass-and-aluminum door. The address is right. There is light inside, but thick meshed curtains block the view of the dining room, predictably not large but not cozy. I turn the knob, open the door, and catch a glance of the almost empty dining room. A man comes eagerly down the central alley to meet me:

"Professor Celada? Please, come in!" and he guides me back to one of the far tables.

Another man, with a short black beard and dark sunglasses sitting on the opposite side of the table, stands up, shakes hands and asks me to please take off my coat and sit in front of him.

In the meantime, my host asks a waiter to bring hors d'oeuvres and then sits down next to his companion. Now both men seem to study my face. The atmosphere is a bit too formal for my taste, but white wine is in the glasses and plates of *Gianchetti* are being served.

Fig.32 - Gianchetti

This is a delicacy for which the Genoese would (and probably will) sacrifice the entire fishing industry of the Ligurian coast: anchovies, caught when they are still two-inch long transparent threads, shocked in boiling water and served cold with lemon and olive oil. I take a mouthful (excellent!) and try to lighten the atmosphere by making conversation:

"Well, what's new in La Spezia? The usual detours on the highway? You said that you had something important to ask me: I am curious, I am all ears!"

Number one (the one with the sunglasses), begins to talk, with an out-of-region, possibly Southern, accent:

"We are sincerely grateful that you could accept our invitation, and I am sure the outcome will be rewarding also for you..."

I take a second mouthful of Gianchetti.

"But first I have something to disclose: you may have noticed a gentleman sitting at the first table near the entrance, with a briefcase on the table… In the briefcase there is a loaded gun…"

I turn my head, and instinctively try to extricate the chair from under the table, but soon give up. The Gianchetti freeze halfway in my throat.

"…and he has orders to use it, should you try anything foolish. Professor Celada, you are in the sights of Prima Linea!"

The late Seventies were Italy's *Anni di Piombo* ("Years of Lead"). The power was in the hands of the Christian Democrats, resilient and impermeable to change, while the Communist Party ran a vocal opposition but was suspected of *Inciucio*, a colorful dialectal expression meaning collusion with the reactionary government to covertly run the country. An entire young generation was frustrated and discouraged. A minority of young intellectuals, some educated in Catholic Universities, some in Marxist-Leninist think tanks, broke off and resorted to violence, united by a poisonous mixture of rage and idealism. They wanted, in their own words, "to strike a deadly blow to the heart of the State". The two foremost organizations were *Brigate Rosse* and *Prima Linea*. They kidnapped, killed or punished judges, politicians, journalists, famously by shooting rounds at knee level. This took place mainly in the Center-North of the country. Prima Linea alone is credited with eleven victims and Genoa, always ready for a social riot, saw more than her share of action. My cousin's brother in law, an inconspicuous industrialist, was among the *Gambizzati*, a word created in those years, to indicate the nasty shooting at the legs.

But me! A newly hired professor of Immunology, outside all power games of the establishment, with progressive views in many directions! I never imagined being at risk of attracting the insurgents' attention and even becoming a candidate for a hold-up. I admit I had acted naively. True, when these people called allegedly from La Spezia, and asked for an appointment to consult with me on a scientific matter that they could not disclose over the phone, I had given the right answer: "By all means, come to the San Martino Hospital: my lab is open!" But when they called again at noon, saying that because of traffic they could not make it in time to the lab, and could I please accept their invitation a little later at a restaurant, I just accepted, meekly! Antonella, my wise wife would have shaken her head and said: "You are too trusting to live in this world".

At the table, I hear myself shout:

"Why me!? What can we possibly have in common?"

Number One answers: "Because, my esteemed friend, we have done our homework. We know a lot about you. We know that you are internationally famous and that and you are the best immunologist (flattery is evidently the second weapon of Prima Linea). We also know where you live in Genoa and that you commute every weekend to your family in Rome, and we have the address of your wife and four children!" (terror is their first weapon). He continues:

"We want your help with a special serum that we need for one of our projects. Of course it is top secret, but you will hear about it soon in the media", he adds proudly. The serum when injected must cause APPARENT DEATH. And we also need the antidote, to enable us to resuscitate the subject. We want you, Professor, to make this serum for us. This project is of utter importance. We know you can do it! You will be compensated for your time. There are no monetary limits."

It takes me ten to fifteen seconds to absorb this speech. The instant they disclosed their identity my brain froze stiff like the Gianchetti. But now, stricken by the absurdity of their design, the brain begins to thaw and to work unexpectedly fast. I can almost hear the frantic "*whirr*" sound of the synapses in action. My gut response is about to erupt:

"My God! You mistake me for Friar Laurence, Romeo's confidant who concocted the potion for Juliet! Welcome to the Sixteenth Century!!"

But my brain has now taken command, and blocks this line from being voiced. Right. Never use irony, and even worse, sarcasm, on people with guns. Instead the brain has already discovered a much better strategy. These guys have mounted an entire revolutionary military action on a lexical error: I must expose the mistake, calmly and clearly. I must be convincing, instead of insulting. It may be difficult, but it is within my professional capacities. I say:

"Gentlemen, you think you need a serum, and, understandably, you call on me because you know that immunologists are experts in serology. However, it appears that what you want is not a serum at all! It is a popular fallacy to call things a wrong name. It happens all the time, and it generates misunderstandings at all levels: with scientific terms the consequences are even worse. For instance, people say *Siero della Verita'* (Truth Serum), to indicate a chemical (*Pentothal*), which decreases the level of self-control: but it is a drug, not a serum!

Number One is surprised, then unhappy. He does not like this line of reasoning and where it may lead. But, alas, he brought it on himself: he has summoned the Professor; now he must endure the lecture. He tries:

"But the Apparent Death …"

I interrupt decisively:

"Look! I cannot tell you whether it exists, but in my entire career I never came across it. And this is why I am the man of SERUM, which is the liquid part of the blood that contains the antibodies. The antibodies are the little molecules that our body, stimulated by vaccinations, manufactures to defend us against infections and diseases, by binding specifically to the invaders, bacteria and viruses and killing them. As you can see, its function is quite different from creating apparent death or other dramatic effects. The substance you seek is not a serum, therefore I am not able to help you …"

"But if it's not a serum, WHAT is it?"

"IF it exists, it may be a drug or the poison of an exotic spider, a toxin of a bacterium… but NOT A SERUM".

I insist on this point because I sense that my kidnappers' conviction begins to crumble.

"Then, who can help us? You must tell me who!"

I play mystified: "IF anybody can, it may be a pharmacologist, or an anesthesiologist, or a toxicologist…"

"But WHO? Give me a name!"

Now I am adamant:

"Sorry, as most researchers in this age of specialization, I concentrate on my own field. And typically ignore the other fields. You have to do your own research: back to the drawing board!"

This is the turning point: they are not able to discuss further. After a long silence, and a quick caucus between them, Number One comes back:

"We will study what you said. You can go now, but we will come back to you, Professor." Then, resuming the menacing tone: "Pray that what you told us is true! And, of course, you must not talk to anybody about this evening. Especially the Police".

He looks me in the eyes now (even though I still do not see his eyes behind the sunglasses):

"Or else…"

I put on my coat and walk to the door. A last look at the infamous briefcase, then out.

Free, out of Piccapietra I go to pick up my car in the parking lot. I drive home in the hills. All is automatic. Not a thought. My brain has taken a bonus leave and for the next two days; my life is purely mechanical. If I do not think, perhaps the Piccapietra event did not occur. On the second day of this limbo, the telephone rings. It is Prima Linea, with a different voice, annoyed:

"Professor, you had been warned! We know you talked to the police!"

Now I am suddenly enraged:

"This is a lie! You are bluffing to scare me. I will not stand for this!" And louder: "Don't dare call me ever again!" and I slam down the receiver. Yes, it has taken place, but now it is out of my system. Directly, I lift the phone again, call Dr. Pericu and make an appointment to see him early next week.

In Rome, that weekend I say nothing to Antonella. I do not want to add any worry to her stress of running a family singlehanded. But I take my son Luca for a walk, the only boy, now eighteen. He is quite impressed by my account. Not only because I add:

"This is a secret between us, that I pass on to you just in case…", but also because 1977 has been wild in the schools of Rome, with demonstrations, occupations and many riots. He has walked quite near the brink of violence: it seemed normal to see *Molotov* cocktails being passed around in the city busses; it was very exciting to be rounded by the police with a group of *Metropolitan Indians* and then be released with the order to get a haircut. Now he realizes what can happen even to his own family when somebody, somewhere, crosses the line. It is sobering. I am glad we had this talk.

Back in Genoa, I go to see Pericu. He is a brilliant lawyer and a colleague, professor at the Law School, on his way to becoming an authority in Administrative Law. He helped me, when I was hired, to dodge problems that may have delayed the assignment of my lab when somebody brought a lawsuit against the hospital and the University. And I trust his political judgment, liberal and realistic. In later years Giuseppe Pericu will serve as Mayor of Genoa for three terms, and both the city and the University will benefit from his efforts. His moment of international fame will come during the G8 meeting of 2001: by a monumental misjudgment of the Secret Services and the police, the center of Genoa will be sealed off with four story-high iron mesh to protect the G8 delegates (among whom G.W. Bush and our newly elected Berlusconi were the most hated by the demonstrators), while the Black Blocks minority will be left free to stage a destructive riot outside the "cage". Cars and *Cassonetti*, the Italian version of dumpsters, will be set on fire, shop windows broken and shattered everywhere. I remember the mayor, standing alone to address the fighting sides with courage and dignity, and finally succeeding in limiting, albeit not avoiding, the tragic consequences, which will include the death of a young demonstrator, and the vicious beating of many others by the police after they were arrested. Like few, I will be proud of being his friend, and, like many, of having voted for him.

At his home, Pericu listens to my story. He is puzzled, not surprised. He is not aware of similar incidents. He thinks the authorities should be informed, but then he says, pondering:

"I believe you, of course, and my advice is to take the matter seriously; but you realize, the only proof of the story is your word and what you remember. There are no phone records. Nobody opened the briefcase and saw the gun. This means that an easy way out for many would be to say it was a hoax, organized to frighten you and maybe convince you to leave your position and go back to Rome".

"Look, Beppe, I wish I could consider it a hoax. I would not leave my chair and run from Genoa, for sure, but it would be easier to digest and classify".

Pericu nods:

"Yes, but one thing I can tell you from experience: in these times there is no easy explanation".

I never heard from Prima Linea again.

Fig.33 - This is the floorplan of the restaurant Il Picchio at the time of my visit to Piccapietra. The drawing is by memory, 50 years after the fact and the restaurant does not exist anymore. It is afternoon, there are no clients, two men invite me at a table, a waiter serves us a plate. Another person sits in a corner, his hands are in a briefcase. Some enlargements are added.

9

Riding with Eli

The Incurable Passion for Searching

Los Angeles. Noon of a luminous day of March 1984. At the end of a sixteen-hour flight from Milano, Franco emerges from the top of the terminal's stairs, and meets his friends, Eli Sercarz and Avrion Mitchison, who have come to pick him up. There is barely time for a big hug and the group leaves the Los Angeles Airport aboard a yellow, well used Volkswagen Minibus, heading towards Park City, Utah.[55] They are going to attend a UCLA conference on the regulation of the Immune Response. Sercarz, quite typically, is behind in the preparation of his paper for the meeting, addressing

55. A version of the trip from Los Angeles to Salt Lake City was dedicated to my late dear friend, colleague and co-father-in-law, Eli Sercarz (1934-2009) and distributed at his memorial in Topanga, LA, on November 30, 2009.

the multiple interactions between T and B lymphocytes, and hopes to use the time of the trip to produce its final version. He is haunted by the term *Immuno-semiotics* that Tomio Tada, the leading immunologist of Japan, has mentioned "en passant" in his last publication as a possible approach to unravel the communication system between lymphocytes. Sercarz senses the importance of the concept and wants to learn more about Semiotics[56]. He needs to get documentation on the subject and plans to read it to his fellow travelers. Unable to locate a textbook in the library, he has borrowed the *Trattato di Semiotica Generale*[57] ("A Theory of Semiotics") by Umberto Eco in Italian, from the bookshelf of Luca Celada, my son, who happened to be a student of Communication at UCLA (Luca is unaware at this time that fate is meditating other involvements for him with the Sercarzs).

Eli looks at me:

"See? Here is the book. It is in Italian, which is a problem for Avrion and me, but luckily you have arrived, and will be our translator-on-board!"

"Of course! But what about the driving?"

"Mark, here, my student from Hawaii, loves to drive. I will pass the wheel on to him when we reach the Valley."

The odds are against the possibility of a serious scientific discussion in the loud van, but the odds are defeated by the unexpected, as always: this time, finding an incredible string of neon "No vacancy" signs on the motels lining the road to Utah. This obliges the three to perform an epic seventeen-hour non-stop drive and gives them ample time for their discussions.

---

56. Semiotics is the science of signs and the basis of any communication.
57. Umberto Eco, *Trattato di Semiotica Generale*, Bompiani Editore, Collana: Studi Bompiani, Milano, 1975.

After the minibus has completed the rising curves to pass the crown of mountains East of LA, the wheel is passed on to Eli's student, and, soon after entering the Fernando Valley, the work starts. Sercarz questioning, Celada frantically reading passages of the text, guessing the meaning of technical terms and translating them into a broken English; Mitchison, scholarly reformulating the issues in real English and acting as a happy father of cell cooperation, the recently discovered phenomenon where two different lymphatic cells embrace specifically and exchange information and orders. Avrion is particularly enthusiastic about Umberto Eco, appreciating both the scholarly and the fictional work: he has just finished reading *The Name of the Rose*, Eco's deconstructed gothic novel, a recent bestseller in England.

In the cozy minibus facts are listed, differences between brain and lymphoid organs stressed, ("But they have the same DNA"! shouts one), speculations are launched about similarities of approach but doubts also are expressed on the compatibility between the experimental method and the procedures of semiotics.

The farther along the road, the deeper we delve into the world of semiotics, and the more familiar we get with its previously irksome language.

"Let antigen[58] be the *Signifier* (originator of information). Where is the *Signified* (recipient of information)"?

"If the T Helper[59] cell is the Signifier, is antigen the Signified"? And who is to interpret the signs? The immunologists? Or, rather, a

58. See *Antigen* definition, Chpt.1, *Porcini*, footnote 4, p.31

59. The *T helper* cell is a type of lymphocyte engaged in the control of the immune defense. They are called "helpers" because without their collaboration the B cells would not carry out the response.

sort of *Brain* within the immune system, made of its components and cells"?

On this last point, a heated discussion flourishes, which is to be carried over to the Salt Lake City workshop. It is clear now, with the benefit of hindsight (in Italian: *Il senno di poi),* that Immunosemioticists prefer the second option: the immune system contains and needs the "talk" among its components. We have full confidence that the immune system is qualified for this job, and it will take full responsibility also for possible mistakes: if there is danger, it will become more efficient; it will broadcast DANGER signals and the entire organism will step up in alarm. Own cells, if infected, will be sacrificed without regret. Bad decisions will be buried with the corpses, good decisions will be incorporated in evolution, and become assets for the future.

At intervals of about four hours it is necessary to refuel and refresh. Both the driver and the minibus are happy to show that, after all those miles of driving in a straight line through the desert, they still remember how to curve, and enter in deserted, almost dilapidated, hot and windy resting areas. A wheel of ripe *Taleggio* cheese that Franco has bought in Milano, following Eli's instructions, before his boarding and smuggled through the customs, provides nourishment during the stops at the service stations in the desert, and contributes to the cozy atmosphere inside the minibus.

Taleggio is a top-quality cheese from the *Po Valley*, white, creamy but solid. It comes in "square wheels", ten inches wide and about two inches thick. At full ripeness the slice has a white dry core, while toward the edges a slightly yellow and soft surface announces an initial melting process. The taste is both mild and sharp, something out of this world. When the travelers cut into wheel, the experience is exactly like the one described above. But at every further opening of the bundle for more tastings the picture changes, the melted portion has grown with the desert heat and soon begins to run. The pungent scent increases exponentially, pleasantly inebriating the travelers. At the end, the affair turns into a race between the eating and the running of the Taleggio. The taste remains excellent and Sercarz, who insists that his family name, in Polish, means "*Cheese-Man*", is elated.

Fig.34 - Eli, Franco and Avrion eat Taleggio in front of their Minibus

The discussion continues. The three men become bolder, as the hours and the miles go by. By the time they arrive at the Great Salt Lake it is apparent that immune cell interaction is a perfect arena for a historic step: the application of the established methods of semiotic analysis to a current problem in biological recognition. The cells of the immune system provide an unparalleled opportunity for the study of biological meaning and of the utilization of signs and signals in regulating intercellular commerce.

Why are these immunologists eager to look outside their own discipline? They are not traitors, nor turncoats. They like immunology a lot. However, if they see any possible connection between their own field and anything interesting outside of it, be it a phenomenon, a process or an idea, their *bloodhound* noses begin to stalk. Nobody can resist testing these possibilities: nothing will stop them from crossing a barrier; certainly not a disciplinary barrier. The latter may be useful, but it is man-made, perhaps old, sometimes ancient, while the object of desire belongs to mother nature, notoriously rude and not giving a damn about imposed rules, but famous for always selecting fitness, and the fit for the future. There is no competition against this choice.

Evolution-supported ways of communications are many, but not infinite and they may partly overlap. The specific immune system[60], a.k.a. *New*, despite being half a billion years old, produces such a large variety of structures (*receptors*[61]), so as to bind and block all possible antigens[62] of all microorganisms, and mounts one type of receptor on each of the circulating lymphocytes. This means that the defense *precedes* the arrival, or even the existence of the invader. B lymphocytes (part of the white blood cells) are those that will respond to an invader by secreting antibodies. But it takes several steps and one alternative:

---

60. This is the new defense style, mounted only in vertebrates, which appeared 500/800 million years ago; it supplemented, rather than replace, the nonspecific defense of earlier multicellular animals.

61. Sites on the membrane of cells dedicated to communication with other cells or signal molecules; all cell collaborations are carried out through sophisticated contacts between receptors.

62. Any molecule that can be bound by an antibody. The smallest unit under this definition is the epitope; only larger antigens are also immunogens, i.e., can provoke antibody response.

1. *B*[63] cell contact the antigen and bind it firmly. This proves that the interaction is specific[64].

2. B cells internalizes and digest the antigens, exposing fragments on their membrane.

3. B cells do not start the full response; they wait for the arrival of *T_H lymphocytes*[65].

4. T_H lymphocytes arrive and bind a piece of antigen (called peptide) exhibited by the B cells.

5. This means: The antigen is "foreign", thus dangerous, and therefore T_H cells diffuse *L2*[66] signaling molecules to start the response.

6. Both B and T_H lymphocytes receive the L2 signal, and begin multiplying and mounting the antibody response[67], while stray L2 molecules excite several inflammatory cells around.

However, there can be an alternative to the critical point 4:

4a. T_H does not arrive: B waits a bit longer, than dies.

The "4a" step happens if: T_H lymphocytes do not exist anymore, having been killed in the thymus, when they were recognized aggressive because they bound a *self-molecule* (a molecule of the same organism) during the routine control that all newly born T cells must undergo. This procedure allows us to write down the two alternative messages handed to the B lymphocytes as:

---

63. The type of lymphocytes that produce antibody responses.
64. Adjective associated to a cell, or an antibody, which means that it will interact only with selected objects, out of many similar ones. Associated to antigen, it means that it will be recognized only by a cell or an antibody that has been produced in an immune response against it.
65. See T Helper definition. Footnote 61, p.159.
66. The most common of small signaling molecules, called *Interleukines*, are able to transmit exciting or repressing signals at short range.
67. The ways in which antibodies win the battle against foreign invaders.

1. "YES, go ahead with the response: you are going to attack an enemy"!

or

2. "NO, stop all procedures, you are a danger and must DIE because the antigen that you would attack is part of our body! There is no pity for potential traitors".

With this list of steps, the three travelers conclude the minibus session with these thoughts:

Avrion: "It was good".

Franco: "We will need a sequel!"

Eli: "But with a Semioticist in the flesh!

Franco: "Why not Umberto Eco? He is in Bologna…"

Eli: "Could we get him? Are you serious?"

Franco: "Leave it to Psmith![68] My former student is in Bologna!"

Two years later, on September 8, 1986. The "Minibus Trio", plus Tomio Tada from Tokyo have organized their dream workshop of *Immuno-Semiotics*. There are sixty participants, including various shades of immunologists, philosophers, mathematicians, with a contingent of semiologists (scientists of sign communication) led by Umberto Eco. Eco, otherwise hard to get because of his growing fame, had been easily contacted through a young faculty colleague of his, Vita Fortunati, the wife of my friend Claudio Franceschi, now chair of Immunology in Bologna.

Eco speaks on day two:

---

68. The masterpiece of Woodhouse is probably *Leave it Psmith*, where the protagonist is a penniless Eton-educated chap, who contrives all sorts of incredible jobs to survive. The first pun of the book is that, whenever questioned, he answers invariably: "My name is Psmith, the "P" is silent", and the educated readers gasp.

"Dear friends, let me say first that I had no idea how complex the Immune System was, and I thank you for the opportunity of learning something new! I don't know whether semiology will be useful to you, but I will help you determine whether our methods to study signs can be applied to your cells. Here is a simple test that we use to probe any potential cognitive system: how does it behave when it receives an *Ambiguous Signal*"?

Several hands rise from the audience, but Eco continues:

"Do you want to hear the definition of ambiguous? I will give you a short message, as an example:

Charles makes love with his wife twice a week; so does Martin.

After a moment of surprise, some laughter is heard…

"As you see, no human mind can tell, from these eleven words whether Charles and Martin woo the same or two different women: It is ambiguous! Therefore, the human mind abstains from an immediate conclusion and waits for, or actively seeks, an independent source to clarify the information."

No more laughter, many smiles.

"By the way, ambiguity is common occurrence, and to dispel it may be vital. Imagine what might happen if Charles or Martin were members of a royal family and a gossip writer should misinterpret the message. Even worst consequences might have loomed if the scene were happening in the Middle Ages"! Eco mimes the act of throat-cutting.

When the hearty laughter subdues, Eco looks at the organizers:

"This is my question to you, immunologists: can you produce an ambiguous signal to test your lymphocytes, or inject an ambiguous vaccine into your mice? You see, responses to "straight signals" may look like the kick you obtain by hitting your knee with a rubber hammer: a mechanical reflex, no food for semiotics!"

It is easy to reassure him on this point: the receptor of the B cell binds a very tiny fraction of the surface of the antigen, called *epitope*, and it is impossible to rule out the existence of a shape similar or identical to that epitope in the entire mouse body. Thus, the response is put on hold until the control is performed. Eco says:

"A control? Does the mouse control itself? Don't you need an independent control?"

Eli intervenes: "Yes! We do! Evolution, with time and diversity has come to this very conclusion: a second type of lymphocyte (T_H cell), instead of recognizing the antigen surface, explores the inside of the same molecule. The B cell will await the arrival of the T_H cell…".

Mitchison takes over: "…You see, when the B cell binds the epitope specifically, it will *engulf* the entire molecule and digest it. Some of the single peptides resulting from the digestion will be exposed on the surface of special receptors (called Class 2[69]). If the T_H cell binds to this "loaded" Class 2, it certifies that the antigen in question is foreign and probably dangerous".

---

69. Class1receptors are on the membrane of all cells of the body. They are showing what the cell is producing to investigative cells (TC) that communicate when the two cells combine their Class 1 receptors. If the TC cell finds something suspicious, like evidence that a virus has entered a cell and taken over the protein-synthesizing machinery to produce more virus cells, the TC kills that cell. To make this specific diagnosis and execution, the TC must be immunized against the virus, which happens during any specific response. Class 2 receptors are only on B cells, T cells and macrophages and serve for communication inside the immune system. It is by an interaction of a T helper cell binding to a Class 2 on the B lymphocytes, that the latter gets permission and a strong encouragement to proceed in its antibody formation.

Umberto Eco, characteristically scratches his head, then with a grimace, scratches a corner of his thick black beard, and begins:

"I am impressed by this step-by-step reasoning inside the body. Particularly, by the *stop-and-wait* time that your cells are willing to take, in order to reach a meaningful, *experimentally correct*, decision... This is definitely not a *tap-and-kick* or a push-button procedure. ...Very interesting. And if I believe what you say (which, given my ignorance in biology, I am happily willing to do)... that the two types of lymphocytes probe different dimensions of the same antigens, fulfilling the requirement of *independent confirmation*... - the audience is still, holding their breath – the Immune System must be considered a COGNITIVE SYSTEM."

Eli: "A Mind?"

Eco: "Yes, primitive perhaps, but a mind!"

The immunologists are radiant, but there are questions.
Mitchison inquires:

"But the immune system is unconscious. Is it a problem for you?"

Eco: "No, because consciousness, as much as we value it, is not a *sine qua non* condition for a system to become a semiology subject". Eco pauses, then resumes: "...and, all things considered, unconsciousness may have a merit: who wants to listen to the conversations of those millions of busybody lymphocytes"?

Fig.35 - Umberto Eco conversing with Sercarz, Celada and Taniguchi at Il Ciocco. This photograph was taken by Rabyn Blake, artist and Eli Sercarz's wife.

Fig.36 - Portrait of Avrion Mitchison, who was present but didn't make in the picture.

Umberto Eco had been drawing doodles during the biologists' presentations. Eco had a witty mind and a good hand. The leukocyte cartoons have adorned the proceedings of the meeting, published in 1988, *The Semiotics of Cellular Communication in the Immune System*[70] and are shown in the next pages:

---

70. *The Semiotics of Cellular Communication in the Immune System* (Spinger-Verlag, Berlin, Heidelberg, New York, 1988), Edited by E. Sercarz, F. Celada, N. Mitchison, T.Tada. The book contains the proceedings of the workshop at *Il*

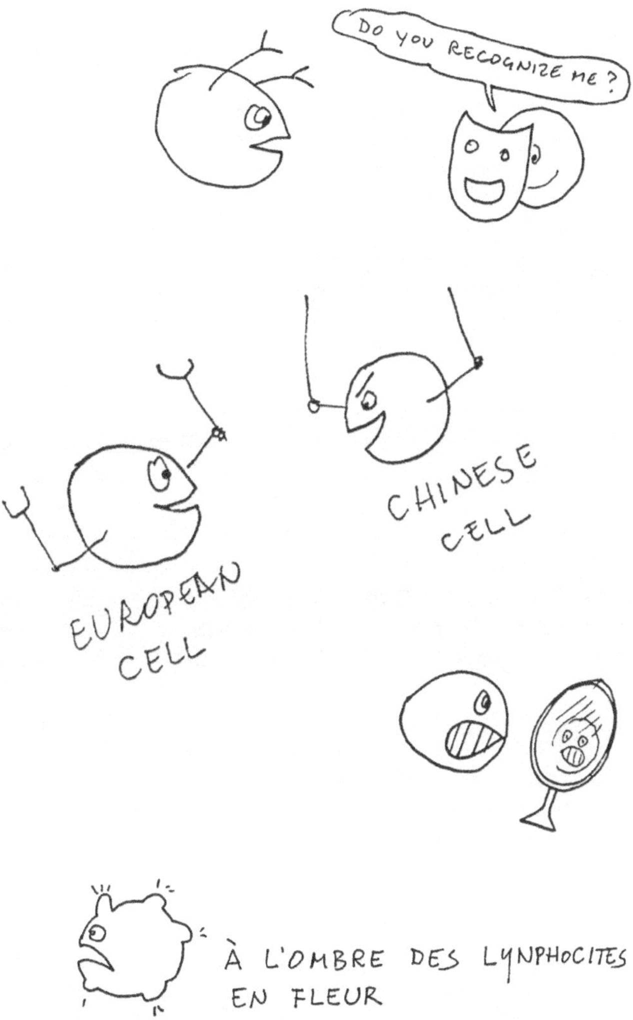

Fig.37-A - Lymphocyte doodles by Umberto Eco. What triggered the wit of a cartoonist at heart? Hidden antigen identity, receptor differences, histocompatibility, the mirror image of an Idio-type and B lymphocyte internalizing and processing an antigen, then displaying peptides on its membrane, waiting for Th cooperation.

Ciocco (1986). The first two introductive essays, of the Department of Semiotics of the University of Bologna, are: *Semiotics and Immunology* by Umberto Eco and *Non-restrictive Semiotics of the Immune System* by Patrizia Violi.

Fig.37/B - Lymphocyte doodles by Umberto Eco: the Idiotypes on display, Macrophages in action, a cell shows an Epitope and an antibody sees that it can bind it!

The immunologists escort Eco to the big blue car and there is a round of handshakes. The car doors are shut, the car begins to move slowly on the crackling gravel, exits the garden gate, the right blinker pops on, the engine emits a little roar and a puff of smoke from two exhaust holes, then disappears, to the right, down the valley.

And now? Avrion, Eli, Franco and Tomio, the four organizers, are pleased to see that the phenomenon of cell cooperation on which all four have produced and published research is being taken seriously by the communication philosophers but they are still hungry: they want to know *what kind* of reasoning the immune cells utilize, to reach decisions like those that were praised in the Garfagnana Valley[71]. They are willing to invent a brand-new name for cell reasoning. Immunologists are masters in naming stuff, but only if they are assured that the same category is not already being used for the human brain. They must find out.

"Hey, the students of Eco had put out a poster about ways of reasoning! Perhaps it is still hanging: let's go back and see! It was illustrated with beans falling from bags, remember?"

We run back, the poster is still there, with three figures, each one illustrating one of the fundamental ways of human reasoning, and getting a grasp at the world around.

---

71. The *Garfagnana* is a valley in Tuscany north of Lucca, where *Il Ciocco* is located.

THE THREE TYPES OF REASONING IN THE HISTORY OF CIVILIZATION

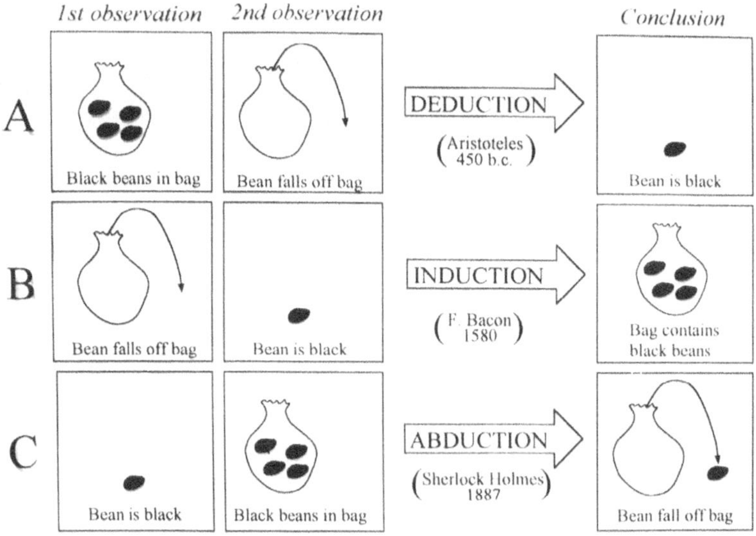

Fig.38 – Examples of reasoning through beans

The first type of reasoning (a), Aristoteles' classical *syllogism*[72], descends from the general to the particular: it is called *Deduction;* if the first two premises are well formulated, the conclusion is inescapable. It is pure logic. However, nothing new is "discovered" through this reasoning.

---

72. In this example, Aristoteles himself wants to certify the concept that Socrates is mortal. He thus constructs a syllogism, composed by three sentences:
1: All men are mortal (Undisputable fact)
2: Socrates is a man (Accepted)
3: Socrates is mortal (Undisputable conclusion)

Deductions like this one are very safe, but they add little new or surprising knowledge

The second (b), the empirical process, ascends from the particular to the general: it is called *Induction*. In our example, all that can be said is that some of the beans in the bag were black – to the limit, only one. The conclusion adds an element of novelty, but it is not infallible.

The third (c), hypothetic reasoning, is called *Abduction* and it is a guess, produced engaging observation and imagination. There is nothing illogical in the hypothesis, but it has no certainty at all; the bean could have fallen from another bag or even directly from the bean plant. Therefore, the conclusion is novel, but it should be tested (corroborated or rejected) by independent information, before being granted the status of "Truth".

The fathers, now orphans, of *Il Ciocco*, have taken a big mouthful, and wonder if all the "styles" of reasoning, as depicted in the cartoons, will possibly be helpful in Immunology.

Franco: "If we need it, we can organize a second get-together with our semioticians."

Eli: "No. We learned a lot: Now it is up to us!"

Eagerly all four jump in.

Tomio: "Syllogism is not for us, and not for the System!"

Avrion: "Isn't *induction* better? Trying to reach a conclusion from observation? After all Galileo was concluding from facts".

Franco: "Can you rule out he was, in fact, making and probing hypotheses? After all, his motto was "*Provando e Riprovando*"! If we translate it as "Testing and Testing again", we can picture Galileo, locating two of his moons and thinking: "I bet the distance between them will decrease during the next hour" (*hypothesis*). Then, after measuring it at different times (*testing*), if the results do not agree, Galileo will make a new hypothesis and a new test, and so on patiently.

Eli: "Wait a minute! Holmes is a researcher of sorts, and seems always busy *deducing*!

Franco: "That is a mistaken word. The detective rather uses induction. Conan Doyle was a good writer and a bad semanticist. There is a book about it[73]. His alleged deductions are in fact *hypotheses*!"

Avrion: "I am relieved, but not too much: I always wondered why his hypotheses always turn out to be right. Mine, for one, are seldom right!"

Franco: "You too would always be right, if your creator were Conan Doyle!"

Eli: "Very funny! Let's go back to the B cells. They are there to defend the body and they must feel in an awkward situation; their weaponry is ready but the actual target could be present in both the invader and the body. And all this happens in a continually changing environment, where both offender and defender may, by chance, mutate. I would use the term *aleatory* to describe this situation. Instead, T cell's "style", to seek an *independent* source of information, is really the best one I can think of!"

Tomio: "Yes I agree! And what you say is present in Eco's test for dissecting *Ambiguity*, and also in the validation of any hypothesis, a.k.a. abduction. I think we are well into *abductive* thinking".

In the Sixteenth Century, Francis Bacon must have realized that Science (the study of Nature) could not go far by proceeding only on syllogisms, and proposed induction as scientific reasoning, putting observations first and trying to reach general principles. This is what

---

73. There have been many studies about the semiology of Holmes. One of them is *The Sign of Three: Dupin, Holmes, Peirce*, edited by Umberto Eco and Thomas Albert Sebeok, published by Indiana University Press (1984).

was taught in Mid-Twentieth Century to high school students worldwide. We were surprised, and felt betrayed, as if a secret had been kept from us for more than 100 years: already in 1888, Charles S. Peirce, in New England, had revealed that the "real" language of Science was *abduction*, proposing hypotheses and routinely validating or negating them. Peirce, now recognized as the father of Pragmatism and Semiotics, and called "The Greatest Modern Philosopher" by Carl Popper, reached notoriety incredibly late. He lived in poverty on a farm in Pennsylvania, haunted by debts, unable to publish his 100.000 pages of manuscripts, and died in 1910 at age 71.

Real friends and Virtual Mind

Avrion: "Will abductive thinking change our way of seeing reality?"

Franco: "Well…, it certainly can change my life, as it encourages me to collaborate with non-biologists. As I told you, I know a physicist at IBM, who makes mathematical models".

Eli: "Yes, you had invited him to come here, but he couldn't come: pity."

Franco: "That's the one: astrophysicist Philip Seiden. He is tremendously curious about the immune system: I explained basic elements of immunology to him and his friends, and they wouldn't stop asking questions. Then he showed me an image on a computer screen and mentioned it was the result of his modeling of the formation of galaxies. It looked to me as one of those NASA photographs, but it was slowly rotating in the cosmos. He wants to come to my lab soon."

Eli: "Why don't you ask him what he thinks of the *Automaton* of our philosopher friend, Varela. His presentation yesterday seemed important, but few of us understood!"

Franco: "Ok, that will be my second question to him and I will let you know!".

1987. New York. NYU. Hospital for Joint Diseases, Sixteenth Floor, my studio.

Philip Seiden is visiting, as he promised.

Franco: "Welcome, Dr. Seiden".

Philip: "Call me Philip"!

Franco: "Ok, Philip, my first question is: having simulated a galaxy, will it be easier to model the Immune Response"?

He is about to answer "no", but I stop him:

"My second question is: What is an AUTOMATON? Francisco Varela, a Brazilian genius from Paris, made a presentation that nobody understood"!

Philip smiles, peacefully: "Automata, invented by another genius, Von Neuman, is a big chunk of a new, but respectable mathematics, called *Discrete,* as opposed to continuous, to which differential equations belong! And, guess what: the galaxy I showed you last time, is technically a *Cellular Automaton*!"

Franco: "Discrete?"

Philip: "Discrete", because it contains a given number of virtual hydrogen atoms distributed randomly in the cells, where, if their concentration reaches a certain breaking point, they coagulate, fuse and generate an immense energy. A star is born by acquiring matter from its surroundings. The difference in concentration varies in waves, following a short list of deterministic laws imposed by the researcher and applied by the computer. Also, time moves in discreet steps: TS1, TS2, TS3… (Time Step 1, 2, 3), like the seconds in an electric clock. And the atoms are distributed in geometric spaces, and move from one space to the adjacent one, according to a project that the researcher writes in the computer code. It is the computer that reads the code, and, at each time step, enacts all the changes. So, we have created time and space and action.

Franco: "I am flabbergasted! But it seems strange that you call cellular something that has nothing to do with cells!"

Philip: "Sorry! My usual forgetfulness when speaking with biologists! Cellular comes from the Latin word "*cella*" and means "room" (as you certainly know). In the model, it indicates the space where the "objects" are contained. For the galaxy model, the rooms are of course three dimensional, but if we begin to simulate the Immune Response, I would opt for 2D, more like a chessboard, or, more precisely, a floor covered with hexagonal tiles, each of which has contact with six more identical tiles.

Franco: "Is there a problem of space? How many cells can we use?

Philip: "There is no limit, but we must begin at a low scale: all the types of lymphocytes you told me about, in low numbers, then we will see; for each type you must tell me all its duties and specificities, and I'll introduce them in the code".

Franco: "All B lymphocytes are identical, but each one has a different receptor, and makes a different antibody. Can you cope with this?

Philip: "Of course: I can use the binary code (a.k.a. bit string), which is a series of zeros and ones. There is no limit to the different antigens or antibodies we can describe, but I suggest we begin with a body of 1,000 spaces and with eight-bit strings".

Franco: "How many antigens can we have"?

Philip: "256"[74]

Franco: "How many different B lymphocytes?"

Philip: "256, but each can multiply and generate…"

Franco: "Clones, of course!"

Philip: "Yes, and when clones are made the antibodies will be secreted, if you want. So, tell me more about how a cell finds its antigen in the real world and what happens after, and I will tell you what can happen *in machina*! But it is a model, therefore it is we, and especially you, the immunologist, who are going to decide and write down the rules!"

Franco: "OK, then this B cell, with its bit string 10010011, is chasing a bacterium tagged 01101100, and needs to enter in contact with this dangerous bug, in order to produce antibodies and beat the infection".

Philip: "Good. Now we have to simulate what you said: physical contact, and binding of antigen and antibody are possible only between entities present in the same space: this is the law I'm writing in our code right now! The computer is going to read the code of behavior of all cells and molecules, and implement the decision for each time step: B cells are stimulated, bacterial antigens are incorporated and, after being digested, a sample of antigen is displayed on another receptor of B cells. However, before we continue, we must decide whether, in this model, there is only binding on the basis of a "perfect match" between antibody and antigen or whether mismatches should be allowed".

74. You can get 256 different binary strings, from 00000000 to 11111111 trying all possible combinations of 0 and 1. There is, of course, a formula for that. The same formula will tell you that, if you use strings with 16bit, the number of combinations rises to 65,536.

Franco: "Phil, we are in biology; there are *high-affinity* (attraction), and *low-affinity* antibodies; also, I like to look at affinity as a "probability of binding".

Phil pauses, then asks:

"How would you like it if we gave a 1.0 probability (a probability of 1, which means a 100% probability) to all perfect matches and a 0.1 probability (which means a 10% probability) to all the cases of one mismatch? It just means that the perfect match will take place as soon as the B cell is in the same space as the bug, while the second mismatch will only have 10% probability to bind immediately".

Franco: "Up to ten times? Will it eventually succeed? Is it just a matter of time?".

Philip: It may succeed or it may not. If not, it will try again in the next time-step. At any additional time step without binding, other probabilities begin their countdown. Every cell or molecule has a given probability to transit into one of the six adjacent spaces. The computer, perfectly impartial, will simply allow chance to present itself by throwing the dice".

Franco: "Phil, I like this system. I think it is "lifelike. But is it a problem to simulate probability?"

Philip: "Good question! And it would have been a problem, if my IBM guys had not invented the way to solve it, by installing Probability Numbers in all computers. I wish I could explain it more accurately to you, but it would take too much time. So, just imagine the computer "rigging"[75] his dices in a way that, if we need any probability, say "0.1", it picks the one that, out of 100 throws, would result in 10 Yes and 90 No, and throws it for you once: whatever the

---

75. Tampering with dice will produce biased results: it is illegal and persecuted. However, when the perpetrators are computer scientists at IBM, they are gladly pardoned.

results, you can trust it, because the machine is correct when tested over 100 throws".

Franco: "Are you saying that the impatient thrower is never right"?

Philip: "He may be, but we are here in a mathematics of "small numbers", where the expected results (e.g. 10%) will be realized exactly only if the dice are thrown, say, one hundred times, or if it is one hundred cells, that throw their dice at the same time steps."

Franco: "How can a lymphocyte throw dices?"

Philip: "It is the computer's task to do that, in accordance with what is written in our code. It is easy since my fellows at IBM have invented a smart way. It's like "rigging" the dice to obtain a given result with the probability chosen by us, which is called bias. The name of this invention is "probability numbers". For us it is very useful. Caveat: do not tell the crooks!"

Franco: "I trust you, Philip. I see a great future working with you and I think this is the beginning of a beautiful friendship![76] How should we call this first probabilistic automaton?"

Philip: "*Immsim*"

Franco: "Good name! It has a nice ring to it!"

Philip: "Well, this calls for a Veuve Clicquot! Cheers!"

Our first serious engagement will be modeling affinity maturation. This means that the average affinity of an immune response increases during the response. A few years earlier, the concept of how maturation happens had triggered a big controversy. Lamarckian and creationists made it an example of specificity molding, or the intervention of a superior intelligence. Darwin

76. This is the last line of the film *Casablanca* (1942), pronounced by Rick Blaine, played by Humphrey Bogart, to Captain Louis Renault.

would explain it as selection of the "fittest" for the job among a large number of responding cells.

In our first engagement, we have 300 B cells and 300 T_H cells, distributed in 16 x 15 linear units = 240 spaces, and with a reasonable probability to move out of their space and into one of the six adjacent spaces.

The experiment begins with the introduction of an excess of antigen. It takes a few time steps (TS) before all cells able to bind perfectly do bind; in this case only one cell binds. It takes, say, 20 TS of moving around of cells and of an antigen to bind also all cells that bind with low affinity: They are 8. Now all 9 "primed" cells must seek the contact of a T_H cell, which will verify that the antigen it bound is legitimate. It takes many TS for all 9 encounters to complete, thus for 9 clones to grow and secrete antibodies.

All this is monitored by the computer, which tells us the average affinity of the primary response[77]: rather low, since only 1 antibody out of 9 has a high affinity.

The *Secondary* (or *Memory*) *Response*[78] will start when all circulating antibodies are gone; then a low dose of antigen is sufficient to challenge most of the high-affinity memory cells, which actually make almost pure high affinity response.

Not only are these data reassuring about the future applications of Immsim, they also announce a prediction[79]: the discriminating factor between primary and secondary response in the selection is the scarcity of T_H cells in the primary compared to the abundance

---

77. Response to an antigen that is contacted for the first time.
78. Response to an antigen that has been contacted and responded to in the past.
79. Results of experiments with mathematical models are called *predictions*. They are considered hypotheses, and need *in vivo* (real life experiments) confirmation to become scientific truth.

in the secondary, where the two cooperating cell "lines" (types) proliferate side by side.

We decide to go public. Our work would fit in a Kingston Winter Symposia meeting in Salt Lake City. The deadline for abstracts has expired but we decide to go anyway. Of course, all sessions are complete.

"Can you present during lunch? It is a plenary slot".

There is a large standing assembly in the lobby, many youngsters holding a half-devoured snack, all eyes pointing to a large screen hanging from the gallery, and, of course, Eli Sercarz, Avrion Mitchison, Paolo Casali, Antonio Lanzavecchia. All seem surprised and positive. It is Avrion who concludes the discussion about Immsim:

"I saw it coming. It looks better than I expected, and we will be expecting a lot"!

1998. Franco says:

"Philip, we must expand Immsim, we have to plan bigger virtual mice with larger numbers of cells for each category. We have to study together the two branches of the immune system: cellular and humoral, which is, by the way, the condition for introducing virus infections and responses. How are your computers? Tired?"

Phil laughs: "My computers are never tired. You know that, when in need, we can always borrow time on the IBM *Supercomputer*."

Franco: "The chess player?"

Philip: "No, his Big Brother![80] I can continue doing it, at no cost, and I will, but in these few years the computer industry is on a surge.

---

80. IBM created a supercomputer that could pre-run and evaluate 20 moves per second. It was called *Deep Blue*, and in 1997 it challenged and beat Gary Kasparov,

Perhaps not far away in the future all we will need is a very thin laptop.

Franco: "If the total repertoire of all specific sub systems is 256 for 8bit strings, with 12bit it becomes 3000, and with 16bit it reaches 56.300! With this large diversity we should be able to achieve much more sophisticated constructions: a good simulation of Nature's almost infinite diversity.

In 1998 Philip and I are invited to participate in a meeting about Interdisciplinary Research organized by the *Center for Interdisciplinary Research*, ZIF (Zentrum fur Interdisziplinare Forshung) of the University of Bielefeld, near the Teutoburger Wald in Germany. It was a preliminary meeting in which ten different groups presented their projects on interdisciplinary biology, in preparation of a conference to be held in Portugal, on Madera Island.

Madera means "timber", reflecting the fact that the island had been a luxuriant forest, even if all trees were cut to make the vessels for the Navigation University of Prince Vincent on Portugal's ocean shore.[81] In 1476, somebody arrived in Madera swimming, escaping from the pirates that killed crews and captured the flotilla of cargo ships of the Genoese Republic, loaded with precious Italian material and directed to Amsterdam. The castaway, by the name of Christopher Columbus, was a true survivor: he did not even try to embark on a ship returning home, where he would face a non-

---

the reigning Chess World Champion. After that, Deep Blue grew to become *Deeper Blue*, but was not allowed to play humans and was retired in a museum.

81. *Escola de Sagres*, near Cabo San Vincente: a real scientific school for discoverers, promoted by the *Infante Enrique* of Portugal, a.k.a. *Henry the Navigator*. Thus, the Portuguese became, with full merit, the most successful navigators all over the world, although they still bear a grudge towards Columbus, whom they had sent packing as incompetent, and Amerigo Vespucci, a passenger on their vessels, whose reports from the New World made him so famous in Europe as to have the New World named after him

sympathetic trial about the loss of his five ships. Instead, he married the daughter of the Governor of Madera, and remained semi-hidden on the Governor's Island, San Pedro. It turned out to be a good place to listen to the stories of passing seamen: a good runner-up university for discoverers, if you couldn't enter Prince Vincent's.

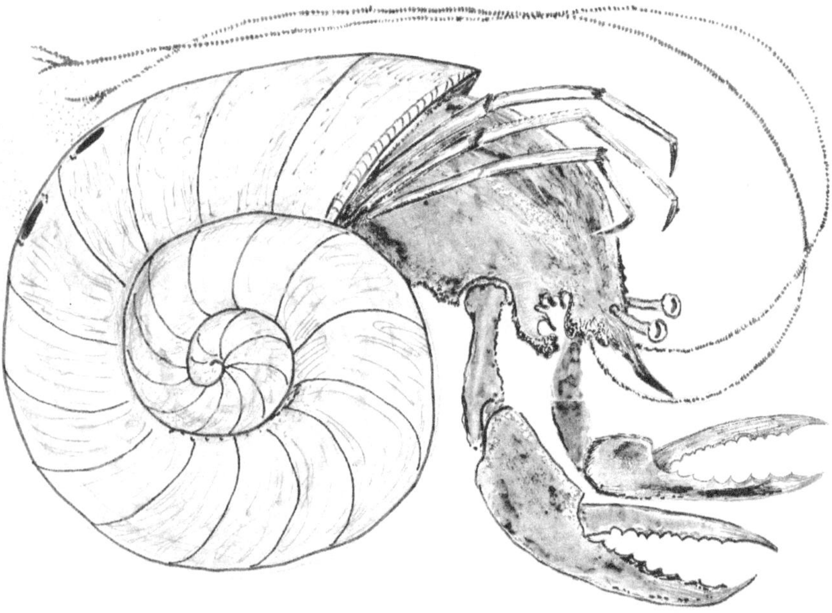

Fig.39 - The one-month encounter of year 2000 in ZIF, at Bielefeld was to promote mathematical models, that is a rare action to decrease complexity in research. We imagined ZIF as monster, roaming in the Teutoburg Forest, where 1,900 years ago the Roman Legions were beaten by the German Barbarians. To stimulate the modelers, I told them I had seen the monster and I produced a picture, colored with coffee and highlighters. He is, as you can see, terrifying, and he is not yet defeated.

In 2000, the great ZIF year begins in Bielefeld, a whole year dedicated to "Inter-disciplinary Research". Immsim is selected for the category of *Computational Modeling* and we are assigned one month to be used ad libitum (encounters, project-writing) provided one week is used for a symposium with fifty people, to promote mathematical models, a rare action to decrease complexity in research. It is a splendid occasion to show and discuss this model, and it's hard to believe that already ten years have passed from my first meeting with Philip.

We imagined ZIF as monster (figure above), roaming in the Teutoburg Forest, where 7,900 years ago the Roman Legions were beaten by the German Barbarians. To stimulate the modelers, I told them I had seen the monster and I produced a picture, colored with coffee and highlighters. He is, as you can see, terrifying, and he is not yet defeated.

Roberto Puzone, a young Italian physicist I had introduced to Philip, is trying to translate Philip's code from an IBM language into a more popular one, called "C". In March, Roberto and I arrive in Bielefeld from Genoa by train and car. The campus is large and green, we like the student-type accommodations. Philip is scheduled to arrive tomorrow.

A phone call from New York reaches me late at night. It is Phil's wife, Lois:

"Franco, Phil cannot come, he's here on the sofa with me, weak and speaking softly, he's not in pain .... We're having problems finding a doctor. Phil asks whether you can suggest a good one. I will drive him wherever!"

"Lois, I will try but it will take several calls and some luck. What is his pulse rate, and blood pressure?

"Pulse is almost 100, but my monitor registers zero pressure."

"He needs a good cardiologist, I'll call you back tomorrow".

It takes many calls from Bielefeld to find a specialist able to handle Philip's heart's failure, in the middle of St. Patrick's festivities, and establish a contact between the doctor's office in New York and the Seidens near Yorktown Heights on the Hudson.

Lois was thankful but, by this time, Phil is even more tired and wants to reach his vacation home in Boca Raton, Florida.

I beg him: "Please, don't; see the doctor first".

Lois is convinced and calls the doctor but can only get the seven days later appointment, despite her insistence. She calls me two hours later from a taxi directed to the airport, having decided to fly to Boca Raton and come back the following week for the visit. I was sad a precious week was lost.

The day of their appointment, around midnight, European time, I could not sleep before calling the cardiologist. He was moderately optimistic:

"Dear Colleague, I saw Mr. Seiden and have reached a diagnosis: your friend has mitral valve failure."

"Is it operable?"

"Yes, as soon the last data come in from the lab, probably tomorrow afternoon"

"In what hospital is he now?"

"He is not hospitalized, I sent him home, and tomorrow..."

"Home!? With a broken valve?!"

"He really needed to rest, and home is best."

Philip Seiden, 67, was taken home and helped to bed. He fell asleep placidly on his side; Lois saw a smile on his face and kissed him on the cheek. The buzzer was near the pillow. She loved this man and had managed to save his life from a previous illness.[82]

---

82. Five years before I met Philip, when he was ravaged by a kidney tumor, the doctors had given up on him and would not try a new therapy from Sweden because its record of cures was too low. Lois, fighting like a lion against all those doctors, argued: "10% ? It is not low! It is infinitely higher than 0! And you cannot

Lois went to sleep downstairs that night, after a taxing, long day. When she walked in his room the next morning, his smile was still there.

April 2000. The ZIF month was ending, and the Immsim group met around a table to see if they still had a future. A respectable number of students, researchers, and professors had been interested, and some were actually working in our (now, sadly, my) groups or in close contact. They were Mathematicians, or MDs with a Mathematical vision, Biologists or students of the two Universities where Immsim was taught: Genova and Princeton. There were Dario Ghersi and Claudia Calcagno, writing their thesis in Genoa, Serenta' and Morpurgo in Milan, Lefevre and Brynja Kohler from Utah in New York, Steven Kleinstein and Martin Weigert in New Haven.

In shock from the loss of Phil, my first instinct at the meeting was to thank everyone, here and elsewhere, and graciously quit.

"Unless one of you, Puzone ... Castiglione... who were consulting with Seiden about his code's translation into the "C" language feels sufficiently advanced to take over his work. In the end it was Roberto Puzone who takes on the challenge.

From ZIF till now, all of Immsim's significant changes had not been technical but strategical, in the choice of phenomena to model and the style to adopt. I met an excellent virologist from Finland[83], who studied anti-virus memory[84] and showed this to be the best observation point for predators chasing their preys, more interesting

---

chose 0, don't you see!?". One doctor agreed. Philip's body weight was below 100 pounds when he received *Interferon*. It was a real miracle cure and Lois truly saved his life.

83. Lisa Selin, Professor at UMASS.

84. A first infection or a vaccine will produce a long lasting capacity by the organism to mount a specific defense.

and less disgusting than those Victorian English shows of Beagles fighting rats. Characteristically one imagines that the virus is the predator; however, after the infection, the roles are swapped and the dangerous predator becomes prey, since the real predator is now the immune system, which remembers the last visit by the same "prey". The race heats up: this predator is ready with his agents, called CT[85], powerful and specific, a masterpiece of selective weaponry. The prey seems hopeless but suddenly, as a true hare, draws its signature side-leap: a "mutation", a minimal genetic change. The mutated virus may be totally unrecognizable, in which case the prey (the virus) wins the match, or not sufficiently different, in which case the predator (the immune system) wins. However, there are a number of intermediate cases. Therefore, Immsim uses systematic modeling to mimic this gray area: for each virus, it creates a virtual series of ten variants, each carrying the minimal difference (one bit) from the preceding in line, and challenges the virtual organism primed (or vaccinated) with the original virus. In this mode, Immsim proved itself to be a sharp investigative tool to dissects what happens in this "grey area". The results are clear, but unexpected: The strength of the response decreases with the antigenic difference, but not proportionally. What happens is that memory keeps its dominance when its anti-virus strength is too low to justify it. This dominance may prove a weakness towards the prey. Immsim predicts that, at a breakpoint, memory will flip from a life-saving tool to a potentially dangerous one. The phenomenon is labeled MaN (Memory anti Naïve) response. Being the product of a model, it has the status of abduction, a.k.a., hypothesis, and will become scientific truth when validated in vivo. In the meantime, the latest Immsim paper, presently becoming ready for publication has found that a natural

85. CT cells are cytotoxic T cells, able to kill infected cells.

effect, aiming at thinning the concentration of cells in lymph nodes, may be a perfect, if aleatory, remedy of MaN. Philip would like this.

Fig.40 - Philip Seiden, brilliant Astrophysicist and Physicist, great colleague and good friend.

10

Stefania

Since I started to write my biographical recollections with the idea of publishing them, my mind has been crowded with so many self-memories and in such variety, that, in order to make a selection, I felt obliged to examine the quality of their contents and how they originated. During this analytical process, one element stood out: the degree of self-identification by the REMEMBERER (ReR) with the REMEMBERED (ReD). I called this property *Self Identification Index,* or "I.I.", which can span from 0 ("I can't believe I have thought, or done, that!" to 100 ("That is exactly my attitude today!"). When applied to all the autobiographical memories, I found the lowest I.I.'s (single digits) in childhood slowly increasing in boyhood, reaching top values in adolescence, and, finally, remaining steady until today.

Unfortunately, even if the number of memories (ReD) is large, the Remember (ReR) is only one, making it impossible to make statistics. However, I challenge people who like to recollect to contact me sharing their experiences. I expect my mailbox to fill, allowing me to calculate the variation of "I.I." growth in different populations, and even test the hypothesis of diagnostic applications in Neuropathology.

For the time being, the effect of applying the "I.I." factor to my self-memories in "Stefania", was the urge to, temporarily, change name to Mimma and Guido's son, from Franco to "Francolino", his nickname as a child. Some friends and family have protested but, since I have the authority here, I'll keep the brat on probation as Francolino, until my identification with him hopefully grows.

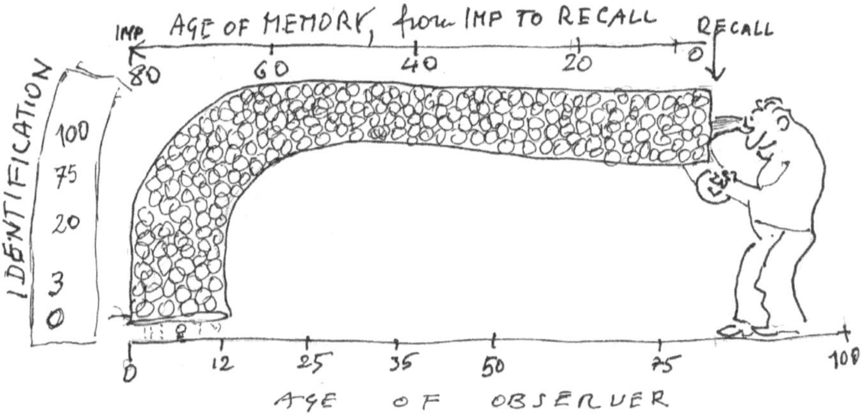

Fig. 41 - The Self-Identificometer, measuring the degree of self-identification by the Rememberer (ReR) with the Remembered (ReD).

Stefania enters this story, revealing her talent in the Olympic specialty of "Brush Pitching", when she begins to throw heavy-duty hair brushes from the fourth-floor balcony to the third-floor balcony of a back-court, typical of the Milano housing style in the thirties. This forces Francolino's mother, Mimma to keep running upstairs with the brushes and Stefania's mother, Alice, to run downstairs apologizing and picking up the next brush. No better way to get acquainted; in fact, a great friendship develops between the two young families: Alice, from Leipzig with Alberto, a Milanese of the outskirts, and Mimma with Guido, married in Genoa and recent immigrants to the big city.

The friendship produces many common events for the couples, such as attendance of operas at La Scala, and, for the entire families, common vacations at the seaside or the hills around the lakes. Soon, in 1931, Francolino is born. His brother, Gianni, will follow four years later, at the same time as Susy, Stefania's sister. Franco has no memories of these early years, but Stefania positively remembers seeing the baby ("really ugly!", she thought) the very day Mimma returns home from the obstetrics ward. It takes four years before Francolino can register memories, and one of his earliest is shared with Stefania.

We are still in Milan, but in a new apartment in Piazza Lavater, where Mimma and Guido just moved. Stefania's family has also moved[86], a block away from the Celadas. A key feature of our home is a corridor, long and wide enough for tri-cycling at ease. The boy and the girl, respectively 4 and 6 years old, are holding their vehicles and, at mid-corridor, they jump on their saddles and start pedaling in opposite directions; at the ends of the corridor they both U-turn with some screeching and squeaking, and speed back. When they cross each other, at the middle point, Francolino shouts:

"Will you marry me?". Stefania answers: "No!"

---

[86] See illustration of *The Moving Party,* p.359

Resuming pedaling that had relented for a moment, the boy and the girl continue their cycling and meet at the end of every loop, coming from opposite sides. Again Francolino, with the same tone of voice shouts the question: "Will you marry me?" and Stefania answers: "No!"

The furious pedaling continues for 3, 4, 5 (or more) times. The result is the same.

Fig. 42, 43 - Franco and Stefania Longoni tricycling

Fig. 44 - Franco proposes

Stefania must have changed her mind about marrying Francolino, between this episode and the next one, which takes place, about a year later, in Magreglio, a resort near Bergamo, with a view of Monte Rosa and the canyon-shaped lake Como. Francolino and Stefania are playing in the steeply sloped garden; it is almost lunchtime, and food will be served on a lower floor of the rented villa.

"We must go down", says Stefania.

Francolino answers:

"I have to pee. I'll be quick. Promise you'll wait for me here, so we go together!"

Stefania: "OK, I'll wait for you."

Francolino goes, relieves himself, then comes back. Stefania is nowhere in sight. (Later investigations establish that Alice has called her to adjust her braids). Francolino erupts in a desperate crying-shouting scene:

"*Mi ha detto una bugia!! Non la sposo piu'!*" ("She lied to me! I won't marry her anymore!")

It takes one hour to calm the boy. Obviously, Stefania is eventually forgiven, as the two remained best friends for many, many years, but Francolino never married her. This memory has a zero I.I. too. The same is true for the next memory, which must have happened shortly after this one.

It is Francolino's birthday, the clue of the party. Besides Stefania, there are two other girls, Enrica (the Brain) and Mariella (the Beauty), also two years older than Francolino. Something falls under the table, where everybody is eating cake, and Francolino goes underneath it to fetch it. He emerges laughing and recites a rhymed couplet that leaves Mimma flabbergasted:

"*Ma che belle cosciotte – hanno queste bambinette!*" ("What nice rounded thighs - these chubby little girls have!").

I hurry to repeat that I.I. of this memory is zero and I am as surprised as I hope the readers will be.

Enrica Tagliabue lives on the third floor of Via Pancaldo, and her family is spending vacation time in Sestri Levante, like us. She will introduce me to the world of cartoons, adventure and fantasy (as reported in Chapter 12, *Sfollati - Part One*) but here she is at the center of an "indirect memory" that her mother confided to Mimma and Mimma, years later, to me. Enrica is four, and writes a diary (a few loose sheets in a folder), which she keeps hidden. Her mother finds it by accident: it contains only pencil scribbles with no order or sense. Mrs. Tagliabue is puzzled. She is about to put back the papers, when Enrica enters the room, sees the trespass and falls in a desperate cry:

Fig.45 - Enrica Tagliabue and her diary

"You stole my secrets! Why did you do it?!"

"My child, I did not see any secrets. There is nothing written there!

"You are lying! I wrote my secrets there, and you stole them!"

Trying to make a logical argument, her mother opens the folder in front of her: "Enrica, let's be reasonable, tell me one single word you wrote in here, and I will admit you are right!"

She thinks she has a clear case, but Enrica cries even more desperately:

"I cannot TELL you what is in the diary, because I cannot READ!

"Then there is nothing to be sorry about! We can be friends again!"

Enrica shakes her head, desolated:

"No!!! I don't know the secrets in the diary, because I cannot read…but YOU can read. Therefore you KNOW them and you STOLE them from me!"

Sestri Levante.[87] On the sandy beach eight-years-old Mariella Calcagni, slender and elegant, with her blond pigtails, is sure to win a "catch me" contest with Francolino, six, who is running after her. The girl already smiles, but at the last turn she loses a few inches, Francolino stretches his hand and loses his balance: while falling, two of his fingers grab the rim of Mariella's red knitted swimsuit, but lose contact again. Both contenders fall in the sand, three meters apart. Rapidly, Mariella jumps up, resumes the run, shouting from afar:

"I win!"

Francolino remains put, sitting in the sand, shocked; and for a reason: while in the air, the red swimsuit had been pulled down a few inches from the suntanned back of Mariella, revealing a white, luminous, sweetly modeled bottom, and was immediately hidden again thanks to the elastic energy of the knitted wool. For Francolino, these seconds of memory have remained a real secret that has not been disclosed until now: 2017. It happened in 1937. It would take seven years before Coppertone advertised its tanning cream using a girl in a swimming suit and a naughty puppy: Decidedly, silly!

Six years old, in 1937 Francolino goes to Primary School. The choice of the German school in Via Boscovich descends directly from the family tree. Grandpa Felice had to learn German the hard way, in order to have access to the Near and Far East markets as a trader. Thus, he sent Guido and Gino to a German School in

---

87. A beautiful resort with sandy beaches at end of the *Tigullio* Gulf, 50 kilometers southeast of Genova. Booming in the '30's, it was particularly beloved by Milanese families with children.

Genova, and they came out fluent in French and English, besides German; Guido obviously followed suit with Francolino and Gianni.

Via Boscovich, which is a rather small Catholic school run by nuns, but more advanced than its Italian counterparts, being a co-ed school. There is a non-confirmed suspicion that these nuns are not liked by the Berlin Government, and for many, this is a plus. I like my nuns, so direct and energetic, even after the "choir-singing accident". The nuns like to host parties during the school term, and are always ready to display the children's choir for any occasion whatsoever. On the first day the choir positions are assigned for the year and Francolino, being short, is placed in the front row, next to Sandro. However, it takes only a few minutes for the sisters to discover that Francolino is hopelessly out of tune. After a moment of panic a decision is made, probably inspired directly by King Solomon: the boy will remain in the front row, lip-sinking the lyrics, but no sound must be emitted, for the next five years.

Francolino always feels at home when Sister Ortensia is in class. She is large, quick-witted, with a gift for miming.

Fig. 46 - Sister Ortensia

Today she explains the most difficult "tense" of verbs, the elusive present tense. The present tense is called *Die Gegenwart* and she pretends to be standing IN FRONT (*gegen* means towards) at the very edge of the platform of a train station where a train slowly approaches from the right and stops.

"In this very moment it waits ("*warten*" means waiting) for me, here. In the next moment the train is gone, "*die Gegenwart*" is gone (*gegangen* is the past tense)! It becomes "*Vergangenheit*", which means "past".

Another day Sister Ortensia explains something totally different:

"Boys and girls, there are things you must learn about your own body: some of its parts are public, everybody can see them; but there are parts, from here (indicating the rope knotted around her waist) and here (indicating the beginning of the legs, under her black tunic), which are "private" and should be kept covered. And for the girls there is an additional "private" zone, from here (the stomach) to here (she indicates just below her neck).

"Any questions?"

There are no questions.

Maria and Sandro Mendini are twins, but the girl is taller. It is easy to become friends with the Mendini because the probability to meet and walk together to school is very high around 8.20 a.m., at the corner of Via Giorgio Jan, which for Francolino is precisely on his way. Contacts are interrupted for a long time because of the war and the choice of different high schools, and then they are resumed. Sandro, calm, always ready with a kind, slightly ironic smile, will make an extraordinary career and business out of his incredible imagination, midway between architecture and design, where nobody can touch him. As a wedding present Mendini will give me and Antonella a dog made of a thousand strikes of different colored pencils. It stays in Tuscany; it likes it there.

Hansi is clearly talented. His mother, Mrs. Fazzari, who married in Italy and remained a widow, has only one mission: to promote Hansi's musical career. He has the "physique du role," screechy voice, a head of red hair, always late, always absent-minded. I run into him years later when I am at the university. He takes me to his home, to his grand piano, his mother in the background. He teaches me something about music, when I ask him how one becomes a great pianist: he plays one note, barely audible, then very lightly, then again and again harder, in precise increments of sound. He says:

"You see? These are twelve steps but, if I want, I can play seventeen, from my, from my *pianissimo* to my *fortissimo*! The best pianist is the one who is able to hit the largest number of increments distinguishable from each other, because he and his piano are the best instrument... I am better at this than most, but mother says I must still improve" and, with both hands, he produces an avalanche of same notes with different strength. He concludes:

"I will wait until they notice me".

I say: "I am sure they will, Hansi! Otherwise, with your talent you could play with any orch…" I am stopped by a screech:

"An orchestra?! Forget it! That is for miserable employees selling their soul for a piece of bread! They will never have me!"

He gets more and more excited:

"You are my friend! You must understand!" He is near, his hands grab my lapels and shakes me… I free myself:

"Hansi! Stop it! I must go now, I'll see you soon!"

I step out of the door, and I am gone.

Back in Milan. It is 1939. Alice's best friend, Erica Colditz comes to visit her from Leipzig, on a short notice. For the occasion Mimma and Guido are invited for dinner. The children go as well, looking forward to the *Strudel*, a specialty of the house; there they will be looked after by the tall and Nordic Fräulein Else. The choice of a German Fräulein, hired through an official agency, is supposed to integrate the school's teaching of the German language, and instill some badly needed discipline into the boys. The evening starts off happily since, for once, the "big ones", Francolino and Stefania (ages 8 and 10), and the "little" ones, Gianni and Susi (4 and 3), seem to have entered a truce. They dine in an adjacent room, connected by an open double door to the dining room where Erica, excited, with her shiny eyes and her dark unkempt curls, is the center of attention. She begins speaking in undertone, but soon loses control of her voice:

"My friends, you have no idea what is happening in our land. They are evil, these Nazis. EVIL! And the biggest ones are in the high places. Our Jewish neighbors have been attacked. Do you know that in Nurnberg all their shops' crystal windows have been shattered? It just happened! Even my church in Leipzig, which is Catholic, has been plundered. Hitler and Goehring want the war at all costs! All Europe is in danger, including Italy!!" She is crying now, desperately.

Alice, emotional about her family in Leipzig, tries to calm her, but fails. The children do not understand what is going on, but sense the gravity of it all; silent, their forks and spoons suspended in mid-air. Suddenly the adults realize that there is a local danger. The two families are not alone. Their eyes turn to Else. She is a piece of ice, her aqua eyes without expression, her lips white for the tension. When the guests go home there may still be some hope for a peaceful ending. When they arrive home the hope is gone. Else is besides herself. She feels guilty and is furious for having been a witness of treason and having done nothing (attack, even kill Erica?) to stop this terrible event.

She shouts:

"*In der Hitlerjugend habe ich gescworen das Vaterland und den Führer mit meinem eigenen Leben zu verteidigen! Gegen Alles! Gegen Alles auf der Welt! Und jezt fuhl ich mich shuldig! Sie müssen mich helfen, das deutsche Konsulat zu erreichen: diese Frau, Colditz, ist eine Verräterin, und muss bestraft werden! Sie sind mein Zeuge!* ("In the Hitlerjugend, I have pledged to defend my Motherland and my Führer with my own life, against all, all in the entire world! And now I feel I am already guilty if I fail to act! You MUST help me to reach the German Consulate! This Colditz woman is a traitor and must be punished!").

Fig. 47 - Fräulein Else: "Heil!"

Looking at Guido:

"And you, Mr. Celada, you are my witness!"

There is a loaded silence.

Mimma puts the children to bed. Guido feels the responsibility towards both Erica and Alice, whom the presence of Else has endangered. It takes him many hours of arguing and explaining, and all his German language skills to calm the girl and avert the incumbent tragedy. He tells me, years later:

"It was my greatest challenge."

After Miss Colditz's prophetic announcement, hints of polarization of spirits and approaching thunders of war begin popping up with increasing frequency.

On Saturday the sound, emitted by a bad, but extremely loud speaker across Piazza Lavater seems more aggressive than usual in summoning young and old to the *Sabato Fascista*, a feast of low quality sports, marches, exercises with fake guns that is supposed to occupy the entire day. Francolino begins to experience the feeling of "non-belonging", or even of "being a misfit", that will mark his entire life.

By pure chance he skips both Fascist and Nazi events because the German School cannot organize Italian gatherings, nor make it obligatory for its pupils to wear or own the *Balilla* uniform required for their Saturdays – the "Fascist Saturdays".

On the other hand, the Italian pupils are not invited to the *Hitlerjugend* exercises, which are obligatory for the German pupils of the German School. The pupils are divided roughly fifty-fifty in nationality. Francolino has the impression that exercises are growing in frequency and intensity. When the big Phys. Ed. teacher, Herr-Sessor Landers, calls, German boys push through the plain clothed Italians, their arms completely bare, their white tank tops deep cut, red-rimmed and bearing a black swastika on the middle of the breast, to reach their leader.

Characteristically, one of the signs of the times noticed by Francolino, with increasing frustration, is the quality of books he is given every term for proficiency. The subject changes according to the student's age, naturally, but also something else changes, and the boy is disappointed. The first year, ending in June 1937, the book was about a school for rabbits in the forest, pleasant, well-illustrated. The second year, it is a book on life in a traditional farm in Pommern (the region of Northern Germany), with a thick straw roof that almost touches the ground, and lots of pigs and piglets. Year three: a book of adventures in Africa where German settlers have to defend themselves (and succeed fabulously) from other settlers who live across a border and happen to be English. The title is: *Der Raub des Dunganen* (Francolino does not understand what this Dungan is, but he supposes it must be some valuable treasure, if the Englishmen are trying so hard to steal it from the good Germans. The book for Year Four is: *Der Skibub vom Adlerpass* ("The Ski-boy from the Eagle Pass"). Here, the handsome 13-years old, Heinz, incredibly skillful on skis, brings messages, saves the life of a deer in distress, but then helps some border SS-guards, who greet him shouting: "*Heil Hitler!*", to capture a foreign spy who tries to penetrate into the Reich.

Until this summer Italy has been untouched by the war, and the family life still follows a peaceful rhythm, even if the storm is all around us. The episode that follows takes place in the summer of 1940. Francolino is 9 years old. Grandma Emilia, Mimma's mother has a large apartment with a large terrace in a resort in the eastern Ligurian Riviera, Sestri Levante. She moves in from Genoa in early June, with her sister, aunt Aurelia, and her daughter aunt Tilde. Mimma's family ("The Milanese") is always invited for the entire summer. Alice's family also opts for Sestri and rents an apartment nearby. They are expected to show up in July. Guido and Alberto commute from Milano every weekend. The children are at the beach every day, but can safely roam the entire village. Today they just came home and are getting ready for dinner. It is June 10th. The Radio is on, a special broadcast has been announced. Mimma and the aunts surround *Nonna* (their grandma), in expectation. The announcer introduces:

"His Excellency, the Chief of Government, Duce of the Fascism, Commander in Chief, Benito Mussolini!"

A wave of noises invades the room as the connection is made with Piazza Venezia in Rome, where the usual oceanic crowd has gathered for the occasion, ready to cheer the Duce's speech at every sentence. Mussolini says:

"*Soldiers! Sailors! Aviators! Black Shirts of the Fascist Revolution! Men and women of Italy and of the Empire! Pay heed*" (Cheers of expectation are growing in the background) "*An hour marked by destiny strikes in the sky of our Motherland…*".

To this "poetry", the crowd answers with extra lively and long cheers. Mussolini waits and savors. Nonna Emilia looks at the aunts, worried, then at Mimma. Francolino sits back, without expression, Gianni, five, is playing with a toy bus in a corner. The Duce resumes where the "poetry" was interrupted, and completes it:

"*… the hour of irrevocable decisions. The war declaration has been delivered to the ambassadors of…*"

Without waiting to hear who the enemies are, the crowd explodes in an applause that evolves into the rhythmic articulation of:

"WAR! WAR!! WAR!!!"

After "long" minutes the Duce can finish his sentence:

"*…of Great Britain and France.*"

In Sestri Levante, the women around the Radio, start crying and hug around grandmother. Gianni continues pushing the toy bus, and emits louder sounds pushing air through pressed lips ("BRRRRR, BRRRR…") to imitate the noise of an engine.

Francolino stands up, approaches the women and says in a steady voice:

"You are a bunch of defeatists!" Then he exits the room.

The women do not even look at the boy, their worry is pervasive. For me, the rememberer of today, it is a shock, and the peak of non-

Fig. 48 - Mimma, Aurelia, Tilde and Emilia crying after Mussolini's declaration of war

identification (I.I. = zero). I was not aware that he knew the meaning of the word defeatist (*disfattista*), a person ready to accept or even bring about defeat, which in 1940 was used in the media and public posters as propaganda against the opponents of the fascist regime.

The little brat is a sponge for what happens around him. Will he ever recover?

The end of Francolino's fifth school year (called *Oberschule Einz* 01), and of his German School altogether, arrives in May 1941. Also this year he has been proficient and his expectations are high. The prize is not a book but a pamphlet containing a black-and-white photo reportage of a field trip, *Hitler In Polen* ("Hitler in Poland"), where the Führer is depicted in every page, riding an open Mercedes, conferring with generals, pointing on the map something they had overlooked, contemplating dead bodies lying on a village square (allegedly shot by retreating Poles), while holding in his arms the child of a German farmer.

Francolino is really disappointed this time. Traces of I.I. begin to appear. Only now do I realize that in '41 the occupation of Poland was a two-and-a-half-year-old news; perhaps Doctor Goebbels was short of more recent successes to keep up the morale of students entering middle school.

11

I.G. Farben

The last year of the Course of Medicine is half consumed. Six years of meeting daily, four hours per day with the same fellow students create a real bond, and often real friendships, especially among those who succeed to stay *on course* by taking all the exams in the allotted times, or share the same turns of the prescribed laboratory and clinical exercises. These routines helped the formation of about 50 boys and girls, out of the initial mass of six hundred freshmen that looked at each other for the first time in November of 1949. Both numbers seam unreal, too large the initial count and strangely few the present graduating fellows. Is it a harsh imposed selection? Or is it natural selection? It is really too early to say: perhaps even half of the group that does not graduate in the summer will pass within six months or one year, and with no harm to their future success and career.

The *nucleus* has several *nucleoli*, and our *nucleolus* is on track, with the exception of Franco, who has decided to do an experimental thesis, after his fellowship at the *Institut Pasteur* in Paris: he plans to graduate in the fall. Some friendships in this group pre-date the start of medical school, and some reach back further, even before high or middle school. It features a future gynecologist, Franco Acerboni, a future internist, Paolo Raineri, a future globetrotting immunologist, yours truly, and a future geriatrician, offspring of a dynasty of directors of private hospitals, Paolo Mombelloni.

"It's kind of sad", meditates Raineri, "though probably inevitable: summer will come and go, and the class will disappear in all directions, like dandelions in the wind ..."

"Yeah", agrees Acerboni, "we should do something together, I mean something notable, to be remembered!".

It is Mombelloni who comes up with the idea of a trip abroad; and, quite typically, moments later, he almost kills it:

"But it must have a catchy destination at no cost, if we want to include the entire graduating contingent: two requirements almost impossible to meet." I come to the rescue:

"Your idea is too good to be discarded without giving it a serious try! Where is Lanzetta? We need his help".

Mombelloni: "Ok, let's go get him!"

We find the future orthopedic surgeon in the plastering lab of the Orthopedic Clinic, in the process of immobilizing the broken leg of a skier. Tall, with a mane, a curved sharp nose, decisive – a real "surgeon mind"- he sees us at the door and waves us to come in, while continuing the plastering.

"What's up? It must be serious, if you come all together like the *Four Musketeers!*"[88]

"We have a job for you: Mombelloni here, has an Idea, but has doubts. We want you to tell us whether it's feasible"! Lanzetta listens and then, with a last, approving smile at the plaster, declares:

---

88. A popular Italian satirical radio show of the 1930's was titled *The Four Musketeers*, mocking Dumas' classic novel *The Three Musketeers*.

"This trip is a very good idea. It can be done. You all know how eager drug companies are to inform doctors about their products. I know there is a race among their reps to secure the first contact with newly-hatched graduates, since they have unbiased, un-imprinted minds and the longest drug-prescribing life expectancy. In a few months we will be highly desirable! Let me just mention that we may consider a graduation trip abroad, and I am sure they will implore us to go visit their plants and be their pampered guests for food, hotel and transportation. Word of Albino Lanzetta!".

In a few days plans become solid. Fifty-four graduating students, thirty-nine boys and fifteen girls, sign in. Our hosts are the three Basel industries, *Hoffman La Roche*, *Ciba* and *Geigy*. At the last moment Lanzetta produces a fourth sponsor from his hat: one of the big German plants, *Hoechst*, which sits on the river Main, not far from Basel. The trip will last six full days.

The large blue bus, fully packed leaves Piazza Castello and takes one hour to reach the Swiss border at Brogeda, but after passing Lugano the speed decreases drastically, when the terrain becomes hilly and we climb and roller-coast towards the Simplon Pass. The smooth tunnels of the future are not yet in place and the curves are a challenge for bus and passengers. The winding rhythm favors social music, and the *Canti della Montagna* ("Mountaineers' Songs") are the first choice for several choral exhibitions, full of enthusiasm, but quite pitiful in quality. A few hours go by. A quartet is modulating the notes of *La morte del Capitano* ("The Death of the Captain")[89] as the bus approaches a lake amidst the mountains.

---

89. Translation of the Italian lyrics: "*The Captain of the company is wounded and about to die. He summons his alpine soldiers and tells them his will: My body in five pieces you are going to cut. The first piece to my sweetheart, to remind her of our first love, the second piece to my mother, the third to my company, the fourth to my fatherland, the last piece to the mountain, she will cover it with roses and flowers.*"

While I consult the road map, something inspires me to grab the microphone, placed in the central corridor, and start the sound-testing:

"1, 2, 3, 4!".

The sound waves interrupt the quartet and infuriates the singers.

"1, 2, 3, 4! Sorry for the intrusion, ladies and gentlemen! This is an announcement! Please pay attention!"

They look up and wonder, while I continue with a fake-solemn tone:

"We are presently entering Canton Uri, the birthplace of Wilhelm Tell and the very cradle of the Swiss Nation! Tell is the marksman forced to shoot an arrow into an apple placed on top of his son's head."

I look at the audience, and sense only moderate interest for history and the events it sparkled. I am obliged to skip the revolt-trigging dialogue between the Austrian Governor and Tell (Question: "Excellent shot! But why did you prepare a second arrow?" Answer: "That was for you, Governor, in case I hurt my boy!") I continue my broadcast, but switch to more practical information:

"On your right you can see Canton Uri bordering of the lake, called The Lake of the Four Cantons (Uri + other three) and in German: *Vierwaldstaettersee* (one word). This is probably one of the longest names of lakes of the German language, but for you, young scholars visiting this German-speaking part of the world, this example is not only interesting, but vital if you want to understand how Germans make words, something that took me five years of German schooling in Via Boscovich 5, Milan!"

I spot a vessel in the middle of the lake:

"Today you are lucky. See that boat, white with the red white-crossed flag, and the smoke line behind it?"

They see.

"It is a steamer and it belongs to the Steamers Company of the Lake of the Four Cantons. In German you use a single word for this Company: *Vierwaldstätterseedampfschiffgesellschaft.* If the company is headed by an Administrative representative (in German, *Vierwalttungsvertreter*), his full title will be, obviously: *Vierwaldstätterseedampfschiffgesellschaftsverwaltungsvertreter.* One word!

Now, I have a little cherry on top of the cake for you, who have followed me so keenly: say this representative owns a pet. Most appropriately it would be a *bassotto*, in English a *German Bassett Hound* and in German, a *Dachshund*, a dog trained to hunt badgers. With this conclusive information and the skills you have acquired, you should now be able to compile by yourselves the full denomination of this particular dog! "..........". Come on! I am sure you have it on the tip of your tongue. Ok, I'll help you: *Vier........wald........ stätter…see…dampfffff…schii ..ff ….*"

I am stopped short! By shouts:

"*Basta!... Buffone!*" ("Stop it, you fool!").

Light objects fly in my direction! I barely dodge a barrage of orange skins! I surrender the microphone and hurry to safety. People become dangerous when they find out they have been taken for a ride!

Besides the goliardic mood of the group, the fifty-four Milanese take in a lot of "Real Switzerland" during this trip. Houses look newly-painted, streets just swept, the grass is trimmed and looks combed. The traffic is extraordinarily respectful. This accumulated perfection does not come as a surprise, as it reinforces the pervading stereotype that Italians have in mind when they say *Svizzera*, which exempts them from admiring the Swiss Model. The Milanese feel particularly antagonized by their neighbors of the North, yet consider themselves superior for a number of reasons: they are the

Metropolis of South Europe, they are traditionally the richest and most industrious Italians. In other words, they are the best and they know it.

Unfortunately, too often does their direct comparison with the Swiss see them failing. The Swiss perfection is objectively hard to beat, and the Milanese assurance makes things worse, since emotion prevents fair evaluation of competitors and generates impatience and eventually sarcasm. This complex feeling reminds me of the dialogue in one of Walt Disney's cartoon albums, between Mickey Mouse and Peg-leg Pete:

Mickey: "See? I was right! Why don't you agree?"

Pete: "Because I hate those who are right! And you are always right!"

The young Milanese, such as our group, are even more reactive toward Swiss perfection than the adults because they are often prone to develop a characteristic form of arrogance, which makes them develop into a "*Bauscia*". Bauscia, in the Milanese dialect means saliva, and the reason for using this term to label those youngsters is anyone's guess. It is assumed that the concentration of "B-men" is significantly higher than average among supporters of one of the local soccer teams (*FC Internazionale*) but this is probably a hoax. In an effort of objectivity, I must admit that the behavior of certain Swiss natives can also be nerve-wreaking, to say the least. I have this recollection, originated in Autumn 1963, seven years in the future compared to the time of the Blue Bus trip.

In a cameo five years in the future, a green Renault 4L is carrying the not-yet-completed globetrotting Celada family on a trip from Torino to Davos, on a high course averaging an altitude of 6,000 feet. The road is a black lucid tape winding under the sun, between three-feet-tall banks of recently plowed snow. Davos is famous for its historical sanatorium, where Thomas Mann set the decaying society of his *Zauberberg* ("Enchanted Mountain"). The same building is now a Microbial Research Laboratory where, looking for a job, I am invited for an interview. The 4L, the only car in sight on the road, rides happily, but suddenly brakes to a stop near the right shoulder: doors open and I come out, clumsily carrying in front of me baby Bebetta in a squatting position, between my extended arms. Bebetta is two, the second child, tiny and funny. A couple of steps and the baby is held high over the snow. Just in time: I resume breathing and I take a moment to look around. Far away, behind the spread-out baby feet, I see a farmhouse, which had remained hidden by the tall bank. A man comes out the door, with long strides and shouts:

"What are you doing!?"

There is no answer: the evidence is clear. A thin yellow hole has perforated the snow under the baby. The farmer is furious:

"This is my property! And this … is Verboten!"

I try to argue:

"Is the snow also yours?" but soon I give up, smile and continue with calculated sweetness:

"Will you please go to hell, sir?"

I turn back to the car with my precious cargo and barely hear the last utterance of the farmer, who has seen the license plate of the 4L:

"Italians! Who else!?".

Sometimes the Swiss ought to be defended, and I am prepared to do it. Do you remember the wonderful movie about the hunt of a culprit by the international police in occupied Vienna, entitled *The Third Man*? The movie is a lesson of morality, but one of the funniest scenes is the outburst by Orson Welles about the respective merits of Italy and Switzerland:

"… In Italy, for thirty years under the Borgias they had warfare, terror, murder, bloodshed. They produced Michelangelo, Da Vinci and the Renaissance. In Switzerland, they had brotherly love, five hundred years of democracy and peace – and what did that produce? The cuckoo clock." My defense at this point has two arguments:

1) Welles, in this film is an embezzler of penicillin (vital and rare at that time) and therefore responsible for the death of hundreds of children. He also mistreats the woman who loves him: he is the villain, and thus not to be believed.

2) The systematic neutrality of Switzerland may well be in part determined by self-interest, but is certainly a real value for the neighboring European nations. Italy, on the other hand, was prone to fall by greed and stupidity into all conflicts of the past centuries. Furthermore, in the moment of truth Italians have found refuge across the Swiss border, and whether they admit it or not, they are profoundly grateful for this.

Back to the Blue Bus trip. After five days in Basel, we are almost tired of visiting pharmaceutical plants. We cross the bridge over the Rhine into Germany following the wide, busy valley heading north toward Frankfurt, riding through several smaller towns. Immediately, something appears to be different both from untouched Switzerland and our home, Lombardy. In the latter, in ten years, the scars of war have been hidden or reconstructed. Here, in Southern Germany and South-West Bavaria, town destruction by war and reconstruction are visible side by side: devastation is massive but the reconstruction of industrial plants and housing is powerful and impressive. We begin to realize that a rational plan has been implemented. The new Germany is building itself like an enormous machine and, only after this phase, will it turn and restore her historical centers, art towns and castles. What we see in Nurnberg confirms this hypothesis; of all the Gothic, steep, arched and colored buildings I had seen in books, and the tight rows of houses bound to each other, only three or four are still standing like desperate trunks after the forest is burned. All the rest are still rubble. They will be reconstructed, probably too well, very soon. Nurnberg is the size of Siena and had been a comparable jewel. To see it in this condition is heartbreaking. Of course we know that war is blind to beauty and other values, and that a particular technique of bombing (*carpet bombing*) has been used systematically by the Allies in order to "break the morale of the enemy" by destroying historical town centers. But here the results are difficult to bear.

In a very short time we reach our destination, Hoechst in Oden Wald, near Frankfurt. We see a pleasant hill on the left, a flat terrain nested on the curve of a meandering river, the Ehrenbach. A tall wall reinforces the river bank; at its end, toward the hill, a massive yellow brick construction is strangely familiar. We have seen it as trademark, on all products of the company, a common sight in the university hospital back at home. It represents a lower-case letter "*h*" (a slanted tower, sustaining a fat arch), which stands for "Hoechst". The bus seems strangely small as it passes under the brick arch. We are inside. The bus parks at the end of a square. On the left, a new glittering office building; on the right, a series of two-story pavilions, occupying in repeated rows of four, the entire walled space; they host laboratories and chemical production machineries, the workspace for thousands of workers. The plant is in full activity, with smoke and vapor, smells and sounds. Industrial vehicles move around.

Fig. 49 - College students visit the Hoechst plant in Germany

We get off the bus and look at the pavilions. At the end of this trip our eyes are well trained to analyze pharmaceutical plants. We are at first perplexed, then astonished. After seeing Nuremberg, we expected to find a completely reconstructed plant, all glass and metal like the Basel companies. But the pavilions are old, and in good shape: their style is pre-war, between 1930 and 1940, and perfectly functional. Are we in a Philip K. Dick novel? Is the blue bus a time-machine in reverse gear? We have too many questions to ask, and to avoid confusion we delegate Raineri, our "square head", to speak for all. Our guide approaches with a large smile, and introduces himself in passable Italian:

"My name is Dr. Otto Brummel, and I welcome you to our Company!"

He is a longtime chemist of Hoechst's who has recently shifted to an administrative post, which is the smart way to reach the top ranks in modern industry. He wants to illustrate the program of our visit. But Raineri, with all of us around him with peeled eyes and ears, speaks up:

"Sir, if you agree, we have a burning question that needs to be answered at once, for our peace of mind: has Hoechst been bombed during the War?"

Brummel signals no.

"No? Then we have a second question: why has Nuremberg been flattened while Hoechst is intact? Was it not part of *I.G. Farben*,[90] which made explosives and gases? Was it not a menace to the Allies, even if they had not heard of the *Zyklon B* gas?"

Raineri's voice is now high pitched, almost trembling. Dr. Brummel is taken by surprise. He wants to calm us, but is honest:

"Dear candidates from Milan, we have many tours coming every week, and you are the first one to ask this. It is a legitimate question and it shows that your heart is in the right place. I wish I could give you a true answer, a real reason, but I don't know it. What I can tell you is that Hoechst has been on the world market for many decades and has collected capitals from international sources. And capitals have no borders."

The fifty-four of us become silent and feel suddenly tired, as if we had grown up too fast. We are about to exit the age of innocence. In adulthood, where we tentatively move our first steps, the air is cool, acts of incredible violence to life and beauty do not make waves nor

---

90 *I.G. Farben* is the short name for the complex of chemical plants operated by the Nazi government to control the production of explosives, poison gas, and all sorts of diabolic concoctions, during WWII. Components of this giant organization were pharmaceutical plants such as Bayer and Hoechst. I.G. Farben was dissolved after undergoing the Nuremberg trials in 1946, but Bayer and Hoechst continued their successful activities until the present time, albeit with new mergers and different names.

raise eyebrows, as long as the motive is presented as "logical": bombing the enemy to save our men's' lives, Carpet bombing to mitigate pilots' losses, burning Dresden as retribution for the destruction of Coventry, or being spared destruction of a plant to avoid a Wall Street monetary loss, like the case we have just witnessed. Today the Milanese candidates did not remain impassible. They are not jaded yet. If this is a trait of youth, I promise to stick to it and never let it go.

12

Sfollati - Part One

October 1942: two years into the war, but surprisingly little impact on the life of the 11-year old Francolino. Last June he has completed his sixth year at the German School and, to start Middle School, he will have to transfer to the Italian *Istituto Gonzaga*, a private all-boys school, run by the Christian Brothers. Mimma and Guido have chosen the conservative bourgeois, solid education for their growing family. The first day of school brings the first surprise, and it shows the nails of the dictatorship that rules the Land. The assembled parents of the new pupils are told that the boys will be incorporated in the Fascist youth organizations and become *Balilla*.[91]

---

91. *Balilla* is the nickname of Giambattista Perasso, a boy who lives in Portoria, a working class neighborhood of the Republic of Genoa in mid-eighteenth century. When some Austrian occupant soldiers threaten to brutalize passers-by, unless they help them extract their mortar from the mud, Balilla shouts: "*Che l'inse*?" "Shall I start it?", and throws a first stone: the following mini-intifada will begin a patriotic revolution. The Fascists, after 1922, used Balilla's name to build their youth organization.

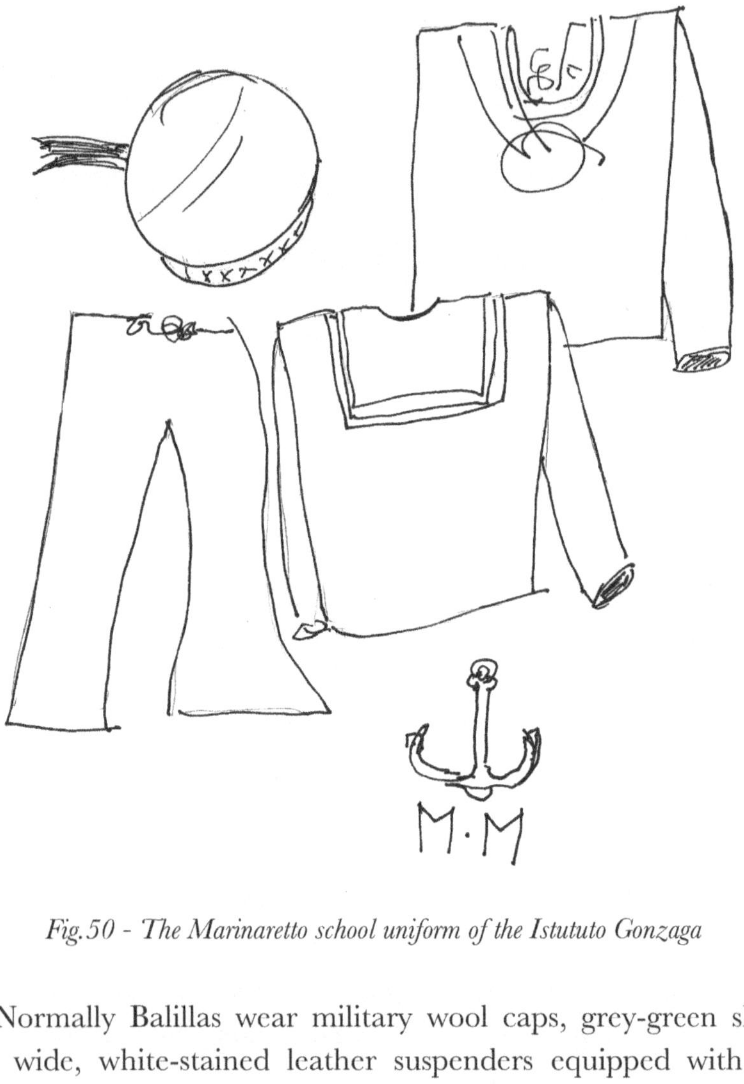

Fig.50 - The Marinaretto school uniform of the Istututo Gonzaga

"Normally Balillas wear military wool caps, grey-green shorts with wide, white-stained leather suspenders equipped with two boxes for ammunitions, black shirts and soldiers' socks and boots". Schoolmaster Brother Clementino, continues with a little smile:

"Admittedly, not too stylish. However, the *Christian Brothers*, who are as patriotic as anybody, have a tradition of excellence and elegance to respect: therefore, we have proposed to host a *special* corps of Balillas in our schools: the *Marinaretti* ("Young Mariners"), inspired, as we are, by the triumphs and the heroic deeds of the Italian Navy while defending *Mare Nostrum* (Latin for "Our Sea") against the Enemy[92]. The idea has been approved by the authorities, and the Brothers are certain that "you, our faithful families will appreciate such a brilliant solution", and will rush to buy the fabric and provide the tailoring of such an elegant uniform!

Sure enough, only a few days later, on October 24, at 4:30 p.m. to be precise, Mimma, aunt Tilde and Francolino ride by tramway to the center of Milano. Under the porches of Corso Vittorio Emanuele II, they enter the store of the *Unione Militare*, a spacious room, dark to our sun trained eyes with thick cords pending from the invisible ceiling. Each cord ends with a factory-style lamp illuminating the top of a large wooden table and the rest of the space seems old, perhaps ancient and shady. Attendants move around balancing thick rolls of fabric on their shoulders, that they open up on the table in front of the costumers with a sudden whirling and thumping move that inundates the desk with dark blue material.

---

92. The performance of the Italian armored ships was disastrous: their enormous cannons could only shoot on sight, while the Royal Navy, hidden behind the horizon, were taking aim at them using their brand new RADAR sensors. The technological gap was enormous and Intelligence should have prevented Mussolini from declaring War. Was the truth kept hidden from him? The Germans had Radar devices, but they were far behind the British level and efficiency in its use: they must have tried and failed to keep this boasting ally out of the battlefield. Successes of the Italian Navy were confined to the small boats and commandos department, where the border between bravery and craziness was often crossed following a tradition started in WWI by commanders Costanzo Ciano and Luigi Rizzo and poet Gabriele D'Annunzio, raiding the Austrian harbors in Istria on board of motorboats armed with one machine gun and two torpedoes. In WW2 the equipment was reduced to the essential: on December 13, 1941, 3 units of two frogmen riding on a large torpedo penetrated into the harbor of Alexandria and sunk several vessels, including the RN Valiant and Queen Elisabeth. An exceptional success, but, still, an exception.

Mimma and Tilde have been shown too many varieties of navy blue to make a quick decision.

Suddenly the Air Alarm sounds! Tooting one, two, three, four, five times. People are shown their way to the *rifugio* (shelter) at least three floors under the street level in the same building of the Unione Militare. The first impressions are lack of light, lack of room and lack of air. Children crying, Aunt Tilde praying……little by little the tension subsides, leaving the tremendous expectation of the unknown. We wait in silence, for almost an hour. When somebody wonders: "is it another false alarm?" and begins to hope, the deep unforgettable sound of the bombs sets off, transmitted, not from above, but through the very bones of Mother Earth. And again. And again. And again.

Then, unexpected silence. A long silence. And, finally, the single, reassuring signal of end of danger. However, our ordeal is not ended: when we emerge on the street it is nine 'o clock a.m. The sky is a river of fire, ambulances scream, we do not recognize our red-lit faces, smoke is all around. On the pavement lies a golden, battered cylinder reflecting glimpses from the sky; the alien object is an incendiary device, now inoffensive, having delivered all its payload of Phosphorus.

Milano is in a state shock after the first air strike, a second one follows six hours later, the attackers take advantage of the still burning fires to find their way. Damages and number of casualties were moderate, yet the Royal Air Force has made its point: their new 73 Lancaster bombers are able to fly from Cornwall to Milano whenever they wish, without Fighter Escort undisturbed by any Italian defense. For the first time RAF display their entire tactical menu: on October 24, in the afternoon they discharge both explosives and incendiary devices on the city, and six hours later they fly in again, comfortably guided by the still lit fires to complete the damage with a second load of regular bombs.

For Italians, *Sfollamento* ("Evacuation") is the only possible answer: it is suggested by the government and planned to the utter detail within each family of the large cities of Northern Italy. Sfollamento means leaving the city, and transfer our lives to the countryside.

For the Milanese, the rational solution would be to move 50 to 70 kilometers north into the *Prealpi* (the foothills of the Alps), the area of the famous lakes and pleasant hills. All these localities can be reached in one hour by a well-developed railway network. This also means that the heads of families could continue to work in Milano, and reunite with their loved ones even every night. These circumstances explain how, within one week of the bombing, 80% of the families with children have left the city.

On October 28, Guido, Mimma, with Francolino and Gianni are on the train to Varese. They have chosen this particular date defying the popular fear that a new air attack would be timed to coincide with the 20th anniversary of Mussolini's *March on Rome*, which marked the beginning of the Fascist Regime.

The train is in fact half empty despite the ongoing rush of *Sfollati*, and they have a comfortable trip. In retrospect it took some nerve by Guido and Mimma to be defiant: this type of word of mouth guesses have rarely been considered truthful, but in time of war acquire some status of popular wisdom, often more reliable than the news on the papers and broadcast by the National radio, totally controlled by the Regime.

While the train whistles and makes way into the increasingly green, pleasant countryside, we are also looking forward to a new chapter of our family friendship with the Longoni. The merit is of grandpa Alfredo Longoni, a shrewd businessman endowed with an above average foresight, who, just before the war, has bought a villa in the Prealpi, on the hunch that his family would need it soon.

Alberto and Alice, have invited the less prudent friends Celada to share Villa Longoni with them:

"You can stay a year, or as long as you please, but we want you to be there before the next Air Strike!".

There are great expectations on the Celada side. Especially Francolino is thrilled, as he needs a change of perspective about the world. Soon after the declaration of war he had begun moving little tricolor tiny flags on the map of North Africa, to keep track of the progress of Italian troops. The experience was heartbreaking, as the flags tended to move in the wrong direction as the Italian Navy and Air force seemed to be unable to keep the promises of the Mussolini propaganda. Slowly and fatally, the little flags were forgotten. A concurrent disturbing element is the growing presence of the German ally, but I believe that the five-year frequentation of the good German nuns of Via Boscovich and of the good German friends helped the 11-year old Francolino avoiding easy, and therefore wrong, generalizations about the people, forces and phenomena that were popping up all around him. The new environment will be interesting; the boy's eyes are wide open.

Villa Longoni. It is a large red brick construction in a tree park, three stories high with a pompous tower growing from a corner and an open terrace topped by a pyramidal roof.

The Celada are assigned quarters and bathroom at the ground floor, near the large kitchen with pantry, while the rest of the floor consists of sale-a-manger, saloon, music room and library. Alberto and Alice with the girls, Stefania, 13 and Susy, 5, occupy the second floor. The Longoni grandparents live on the third floor: They have the best view, but the three flights of stairs limit their mobility, and they are seen only at dinner. The villa is certainly impressive, but, at the first inspection, Stefania and Francolino are attracted by two features which no adult would have considered worth of attention. One is the basement, accessible from the regular stairs and from the garden, through a glass walled Green House. The inner basement contains a henhouse, a row of tall rabbit cages and a large laundry room. The entire area is under the jurisdiction of Marcella, the rustic cook from Mantova, and factotum of the Longoni family. The second, and for us more significant attraction is a particular assemblage of tall trees and thick bushes in a corner of the park, forming a natural "house in the woods" with several "floors" reachable by easy climbs, many secluded spots, and openings to a turgid lawn. From the first moment Stefania and Francolino know that they have found their own, exclusive place, where they can hide from the world and live their exotic adventures. They are both surprised at the suddenness and power of this fantasy. This must have been there before, but kept buried in the unconscious until an ideal environment permitted its eruption. The inspiration comes from movies, and especially from American cartoons, that in the late Thirties have been imported, translated in Italian and published by several weekly papers "for boys". *L'Uomo Mascherato* ("The Shadow"), *Giorgio Ventura* ("Brick Bradford"), *Cino e Franco* ("Tyler & Spud") were mindboggling, sadly missed since the War outbreak, and, in this case, they became the subject of half-told/half-acted adventures. Most of them are exotic, as they often take place in the Jungle, all of them are thrilling, even heroic. Many feature animals: black panthers, galloping horses, buffalos, etc. A recurring mission is to protect elephants from poachers and ivory merchants. Stefania is adamant:

"For all our tasks in the jungle and the forest hideouts out, we need to have secret identities, right?

Francolino is thrilled; he nods in agreement.

"Then, I will be Marco, a strong, tall young man, and you must choose your own secret name, as another young man, a little younger, a little shorter…". She looks at him for a while…"but very fast and a hell of a swimmer!" Francolino says:

"Ok, I will be Giorgio[93]. We will be the best of teams".

Looking back, Stefania's speech, and Francolino's response come as an unexpected and drastic change of attitude and feeling in both characters: only last year (see Chapter 10, *Stefania*) their dialogue and obsession was on whether they would marry or not, about which they were always in disagreement. Now they are busy with a war and their fantasies, always the best of friends, this time, the two "boys" are too focused on their adventures - not to mention the war - to even think of love related matters.

The two "young men" are slender, agile, able in archery and shooting, as the last resort. There are enemies to be avoided and lives to be saved. It is imperative to hide from the *Pigmies*, a wild African Forest tribe dominated by *The Shadow*, which are represented here by the *Little Ones*, Susy and Gianni, always insufferable, and always ready to intrude into our habitat, if not into our fantasies. Another tribe to hide from and spy on are the *Zulu*, who live next door, allegedly as sons of the farmer that supplied fresh milk to the Sfollati: they become mean and out of control when the Shadow is away on some perilous mission, and try to hit us with stones, a task fortunately nullified by the thick foliage of our trees.

93. *Giorgio Ventura* is the Italian name for *Brick Bradford*, a science fiction comic strip created by writer William Ritt and artist Clarence Gray.

The third dangerous being to tackle is of course Marcella, but she is assigned, case by case, a variety of roles, depending of the story she happens to disturb (I wonder what she would think if she knew about her many "involuntary" impersonations).

It is not easy to understand the role of the rabbits in these fantasies: they are loved, picked up from the cages by the skin of their neck (NEVER by their ears!) manipulated, subdued and even hypnotized. They have names: the two first of a dynasty are Hansel and Gretelinde. Gretelinde is all-white, while Hansel wears pants: his rear is solid-black, and his front half white. The rabbits are not given specific fantasy roles, but handling them fulfills a need evidently important in adolescence, especially for City-raised children. They have also a role in our physical training. Often we set the rabbits free on the lawn were they begin immediately to eat grass meticulously and professionally. They are tame and in the first minutes of freedom it is easy to pick them up again. But if they are allowed to stay free a little longer, the pick-up operation is progressively more difficult, on the verge of an athletic feat: freedom is addictive, rabbits become hares. This means that after one hour we have to dash like crazy soccer goal-keepers in order to secure the jumping beasts before 7 p.m. to avoid penalty. Penalty means no rabbits for two days, no appeal and Alice is a severe enforcer of all rules.

The first months of Sfollamento, chaotic and disorganized as they come, offer contacts with many children and families in a similar condition of forced *Villeggiatura* (vacation).

The City of Varese has set up several middle school classes in the Villa Reale in the middle of the town's public park. The funny white tramway cars used to commute to school are always packed with loud Milanese students, who (like the rabbits on the lawn) experience a newfound freedom and feel and act like the owners of the town.

The school itself and, particularly the teaching, is a disaster, it must have been organized at the last minute, with all teachers being temporary and continuously replaced. Class-1, where Francolino sits will have changed seven teachers by June. Imagine starting Latin in these conditions!

Thin Carlo Vandoni, of 1-F and offspring of friends of the family, pretends to mistake *Latino* ("Latin") with *lattino* ("sweet milk") and comes up with a pitiful joke:

"I hope it tastes good!"

The parents are worried about our education. Somebody proposes a committee to meet the Schoolmaster, which would be a wise initiative in a reasonable society. Nothing is done about it, the society wasn't reasonable enough.

A night in late November, everyone's secret fear materializes. It is midnight, and a sound similar to a thunderstorm, far and persistent, compels us to go up to the tower. Straight South, Milano is under air strike.

Pulsing lights, in groups illuminate the sky at the horizon, the attack lasts half an hour, but it will be repeated once, twice, three times per week, with more and more bombs as part of the carpet bombing strategy and the rush to the tower will become a routine for the Sfollati, who are learning how to live with the worst.

After a still peace-like Christmas, with *Oh, Tannenbaum, Oh Tannenbaum, wie gruen sind deine Blaetter* sung by Alice at the piano and accompanied by the children, come two symptoms of harshness: the scarce, "spartan" presents under the Christmas tree and the replacement of Alice's signature *Pfefferkuchen* and *Krapfen* with a thrifty bread pudding, which allowed the re-use of stale bread. For the record, Francolino finds the pudding delicious, and asks for a second serving. This obliges Stefania to follow suite, against the better judgment of her stomach, and she gets sick.

Fighting on Italian soil begins on June 10, 1943 with the fall of *Pantelleria* and *Lampedusa*, two rocky islands sitting mid-way between Africa and Sicily.

In a broadcast released from Algiers, Churchill and Roosevelt warn Italians of the imminent attack and ask them whether they want to die for the Duce and the Führer or live for Italy and a future of Liberty.

Mussolini reacts scornfully on June 24 declaring that should the Allies attempt a landing in Sicily, they will be stopped at the water's edge.[94]

On July 10, in spite of a prohibitive stormy weather (with Force 7 waves), the landing begins in Sicily and the next day American and British forces walk into *Gela* and *Siracusa*. Nobody stops them, and they look unstoppable.

The Axis forces are strong in numbers, but the Germans, in arrogant distrust of the Italians, have taken over the command and decide to save their divisions with a quick retreat.

The Sfollati, against odds, are well abreast of the developing news. Everybody listens to the *Radio Londra* broadcasts, preceded by the deep sound echoing the incipit of Beethoven's Fifth Symphony: "Bam-Bam-Bam - BAM!", which is still in the memory of an entire generation. Another source of free news was the *Corriere del Ticino*, a Swiss newspaper from an Italian speaking Canton, which is regularly smuggled in from the nearest border.

---

94. This declaration became known as the "Water's Edge Speech" (*Disorso del Bagnasciuga*), one of the many arrogant statements of Mussolini that did not come true.

At Villa Longoni, the Sfollati are overwhelmed and react in different ways.

Alberto is gloomy and everyone else feels that the war is lost and is prepared, or even eager, to accept the inevitable without qualms. But the wheel has not reached a resting point.

The diminutive King Vittorio Emanuele III, labeled, with some irony, the Soldier King is still the commander in chief and gets cold feet. He is responsible for putting Mussolini in power in 1922 and accepting from him the title of Emperor in 1935, after a shameful war in Ethiopia. He signed Mussolini's racist laws in 1938 but now he is scheming against him to save his pants and replace him. A handful of fascist chiefs have similar thoughts.

On July 25, at the Annual Great Council of his own party they come out with a nicely worded resolution, praising the Duce, but asking him to step aside, thereby allowing a better way to end the war and the sufferance of the Italian people. Among the signers is Galeazzo Ciano, Mussolini's son-in-law.

The resolution passes, totally surprising the Duce, who is obliged to report to the King and still counts on His Majesty to solve the misunderstanding of the Council's vote. The King asks and immediately accepts his resignation as Prime Minister. Mussolini begins to smell something foul. The King cordially accompanies him to the door...and, a few steps outside the Royal Palace, gets him arrested by a Colonel of the Carabinieri and sends him to a prison on the impervious Gran Sasso, on top of the Apennine Mountains.

The news is a shock for all Italians, who are sharply divided and see themselves either freed from the oppressive Regime or in imminent danger. The Germans are furious and begin to plan the full take-over of Italy. However, the sequence of actions and news are confused, contradictory and ridiculous.

The King asks a General with a dubious reputation in three wars to take over: General Badoglio.[95] In his first address to the nation, he declares: "The War continues", leaving the Italians, the Allies and the Sfollati) puzzled. It will take a month and a half of operetta-style contacts with the Allies (through envoys that did not speak English and traveled six days by train to Lisbon) to sign Italy's capitulation while keeping it secret from the Germans.

At Villa Longoni, Francolino overhears loud sounds coming from the library's half-open door. It is Alberto shouting:

"Traitors! Cowards! That is what we are! A dishonored Nation that walks out on its Ally without having the guts to tell them!"

There is a second voice coming from the library, subdued and sad. Guido replies:

"I can't take it anymore! I can't live with my family in a house where our countrymen are called traitors. Entering the war was a mistake and a violence against the Italian people, which fatally ended in an act of self-defense."

There is a silence. Francolino, in the next room is confusedly proud of Guido speaking up. Alberto realizes that he has gone too far and changes tone:

"Dear Guido I did not mean it this way! I know well that in the Great War you were one of the real brave officers: you enrolled as a volunteer, not once, but twice![96] I will forever admire you for your courage. It is the lack of decency of our rulers that infuriates me!"

Francolino does not know whether the two men embraced after this exchange, but in this precise moment he feels that the two-family

---

95. Pietro Badoglio was Commander of the section of the front at Caporetto where, in 1917, the German Straf-expedition broke through with unexpected ease.

96. Guido Celada volunteered at the beginning of World War I, was wounded in the foot in 1916 and honorably discharged. In 1918, after the Caporetto defeat, he volunteered again and served as a lieutenant until the end of the war.

community comes to an end. He gasps, as one who receives a blow in the stomach. He also learns a truth of life: impulsive Alberto, by marrying a German, has become much more pro-German than sweet and rational Alice will ever be.

A little consolation: the bonds between the mothers and the children are not in the least affected and will last for life, impermeable to political waves.

The next morning Mimma, petite, piercing bright eyes, wearing a triangular felt hat with the feather tilted to the right, takes the trolley to the Housing Office for Sfollati. Long waiting lines. Confusion. No apartments available.

Mimma forces herself through an office door and gets the attention of a senior Civil Servant, who tries to discourage her:

"You see, Lady, the first wave of requisitions of apartments is over, and the owners have become more difficult…for instance one of them, an important name (perhaps even a nobleman) will oppose our requisition if the prospective tenant does not show three references, obviously, from important people. They will even refuse to show me the building: a hopeless situation. You better give up!"

Mr. Ghiringhelli is puzzled to see that every argument he uses to get rid of this Milanese becomes fuel for Mimma's determination:

"Three References!? I will get you six, and from the best sources! Distance: I am not afraid of that. There are trams! The building is ancient? Difficult to clean? Full of bookshelves? My boys are respectful! There will be no complaints from the owner! You are a smart and kind person Mr. Ghiringhelli. All you need to do is introduce me to this Doctor, Professor… Nobleman, whatever he is and I guarantee you will earn his praise and gratitude and, of course, my own!"

Ghiringhelli is overwhelmed by the lady's determination (he does not know that Mimma's nickname on Tennis Courts used to be

Granin di pepe, "Pepper Seed"). When he tries to reply he is mesmerized by her piercing eyes and vibrating feather...and realizes he cannot win. He falls back in his chair and grabs the telephone.

Mimma understands she must step out of the office and wait outside. Later, the door opens again:

"You have an appointment tomorrow at 11, which does not mean you are accepted".

The next morning, she wears her best tailleur, a sedate hat with a veil and finds Francolino:

"Get ready, I must take you with me to the appointment, since father is in Milano! And you better behave"!

The boy is proud to be asked and proud to have such a strong and brave mother.

Velate, as the crow flies, is perhaps one kilometer from S. Ambrogio and two hundred meters higher in altitude and the road winding around hills with sharp curves and sudden climbs makes the trolley car look like a toy. They reach the church in forty minutes. The church sits on top of the hill surrounded by the village dwellings, tiny, irregular, conglomerated as if seeking protection.

The hill is so steep that the traditional square in front of the church (*Piazza della Chiesa*) is replaced by a long, straight stairway that stretches from the church door to the village main street.

Directly across the Church, stands a large iron gate that closes the access to a private park, but cannot hide the richness of secular and exotic trees it contains: it is majestic. Grasping the gate's iron bars and poking his head through them, Francolino tries to see more but Mimma wants him to help her find the address she is seeking. Riding a streak of luck, they find the address on the rustic wooden door on the wall right next to the park gate.

Fig.51 - Mimma and Francolino search for a new home

13

Sfollati - Part two

Two colonnaded courts: the first, accessed through the gatehouse, has covered walkways, with a farmer residence, animal quarters, working tools and gear.

A tall arch carved in a self-standing Ivy-covered wall opens to the white graveled plaza of the Manor's residence. It consists of two bodies of masonry joined at L and protected by a wide *Loggia,* embracing the plaza and ending with a sturdy tower overlooking the park. The Manor has two dwelling levels; large paneled windows elegantly spaced on the main floor, and smaller ones under the roof to give light to the traditional attic.

On the ground floor, the colonnaded walkway framed by slender round arches, unmistakably Renaissance style, takes half of the space.

Behind this gallery, runs a sequence of rooms, originally assigned to education and entertainment: libraries, sitting rooms, music rooms, testify to the pleasant life of the cultivated wealthy.

Fig.52(a) - Aerial view of Villa Clerici

The landlord's family living quarters occupy the ample main floor. Mimma and Francolino stare at each other in amazement, when a group of loud children come out running under the porch, heading towards the trees.

"... My grandchildren! They are wild..." a man complains, shaking his head. It is Mr. Emilio Clerici, a frail gentleman with a walking cane, who approaches us and continues, addressing Mimma:

"Madame, my daughter Giulia tells me that you are a friend of friends of hers: For me, this is more valuable than any letter, especially in times like these. I will notify the Town office that you are accepted, and I hope, that, even if you arrived here by pure hazard, you will find a livable accommodation within our rural estate!"

Mimma thinks: "Have we come to an oasis, outside the war?" She is radiant: not only is she at ease in this mostly Milanese society, but she seems to accept all this as a gracious gift between members of the same social group, and even Francolino knows that the price was paid in full. A section of the ground floor, that reaches the tower, has been separated from the rest of the villa by obliterating one door inside the apartment and by raising a wall - to separate the last section of the colonnaded walkway.

Fig.52(b) - Villa Clerici, view of the tower

The concrete wall across the porch is a bit shocking, but these Sfollati realize the privacy gained: a large part of the activities will be carried out in this section of porch, and nobody, unless he or she comes walking a long way on the gravel, can see it. Inside there are two bedrooms, a large kitchen with wood burning stove and a bathroom, tiny and funny shaped, having been adapted from an under-stair cabinet of the tower.

In the old times the boys' bedroom had been part of the Manor's library, and contains many vestiges and still displays its furniture and its contents: Two identical bookcases, eight row tall and glass-paneled, that together occupy the back wall of the room; the pieces of furniture themselves are Eighteen century Antiques, and the contents are a complete collection of LA REVUE des DEUX MONDES, the periodical magazine, which published the most significant works of French Nineteenth Century literature as serials.

The two boys are too young and too busy to try to read these treasures, but the keys to the showcases were in the locks. When alone, Francolino learns how to slip out a volume from the lower rows, gently browse through the pages, often uncut, and put it back where it belongs. All are paperbacks in excellent condition, with a kaki/peach cover and an antique scent that seems to climb up to the brain…the boy is aware of the exceptionality of being in this room and of holding in his hands these 100 years old volumes…

A note by the author:

Why do I still call the protagonist Francolino? He is 12; the way he followed Mimma in this particular mission; the very way he is attracted by the rooms and the arches and especially by the lab are very similar to the way of thinking of the mature version of his persona. From this point on, I will risk calling him Franco, ad letting him be me, at least on probation. The boy is not expected to be reasonable all the time, but, should he act stupidly or be unable to resist nonsense, he must be ready to take responsibility for his actions.

The converted apartment is the most positive family news of this autumn but the Sfollati realize that they are like snowflakes in a slow-moving avalanche, which has just shown a number of abrupt accelerations.

Late 1943. Italy has tried to exit the war, but war has remained around and inside it. The Southern third is liberated by the Allies, the remaining two thirds are in German hands. The Italian army is dissolved: few corps have made it to join the Allies. The majority has abandoned the arms and try to hide at home; some have joined rebel groups fighting the Germans, which are piecemeal gaining the upper hand; others have been captured by the Germans and sent to Working Camps in the extended Germany or are facing a tragic, slow destruction as ARMIR, a corps of 230,000 sent to fight an impossible war in Russia, the most painful toll imposed by Hitler to Mussolini, and finally paid in terms of 110,000 casualties by the Italian people. The strength of the Partisans is to have a unified commanding organization, CNL (*National Committee of Liberation*), supported and armed by the Allies. Their weakness is to be divided into groups with three distinct ideological motivations. The ideology is immediately recognized by the color of a scarf worn around the neck, red, blue or tri-color, like the national flag, respectively coding Communist, Catholic or Liberal.

Ideologies, like religions, always end up being divisive, despite all heralded declarations and convictions of the contrary. A new world is emerging after the long domination, but the butterfly creeping out of this larva is still not imaginable.

Rumor has it that Hitler has sent a *Storch*[97] plane flown by the daredevil Captain Skorzeny, to smuggle Mussolini out of prison on

---

[97] *The Fieseler Fi 156 Storch* (English: *Stork*) was a small German liaison aircraft built by Fieseler before and during World War II. It remains famous for its excellent STOL (short take-off and landing) performance.

top of the impervious *Gran Sasso* (where the King has forgotten him when he made a dash to Puglia!)[98].

Fig.53 - The Gran Sasso raid: a Storch plane smuggles Mussolini from his prison at Campo Imperatore.

They succeed. When the *Storch* lands in Munich with Mussolini, Hitler hugs his old friend and assures him that he wants to support

[98] See illustration at p. 358

his wonderful idea of creating a Neo-fascist Republic in Northern Italy. The translation of this double talk is the following:

"I saved your life. Now, you must work for me whether you like it or not. There is another thing: the traitors who voted against you will not be forgiven." The Duce thinks of Ciano, father of his grandchildren and probes:

"Are you going to punish them?"

"No, *you* will. After the trial, with a shot in the back!"[99]

All this is part of the German defense plan, meticulously applied, which succeeds in blocking the Allies for almost two years. The clock slows down suddenly. War is prolonged and Sfollamento seems to become, if not a desirable way of life, at least, livable.

Velate is a world to explore, and the Manor is their first objective of exploration. However, can newcomers explore without the help of native guides? Fortunately, the Manor itself offers insiders: the four grandchildren, offspring of Giulia and her Swiss architect husband, Fabio Franchini, and their ages seem handpicked: two girls, Franca,13, Giovanna, 8, have ages well matched with Franco and Gianni. Three more boys, Alberto, 10 and two younger ones, Luca and Paolo.

Franca Franchini, always game for a new proposal, frizzy blondish hair and an explosive laugh, gets the immediate admiration

---

[99] The trial, ordered by the newborn Italian Social Republic took place in Verona, Jan 8-10, 1944; six of the *Gerarchi* (hierarchs), who had voted to replace Mussolini at the *Gran Consiglio* of July 25,1943, were found guilty of treason and five of them, including Count Galeazzo Ciano, were shot on January 11, 1944.

of Franco, even before finding out whether she will support all his projects, no matter how crazy. Will she?

Fig.54 - Franca Franchini on her bike

Perhaps, but only after having examined them with the full authority and experience of the "one-year-older".

Today Franca shows up with a ring of dangling iron keys around her wrist:

"We can begin by exploring the attics".

On the second floor, she chooses one key, by the attached cardboard label, and opens the door to a steep stairway to the attic. At the top they see the structure of a cathedral roof. The majestic structure, sustaining the red brick roof, is made of roughly cut pine trunks running from the central to the external wall, and cover an ample space. Franca opens a series of doors that lead to rooms with slanted ceilings and wide oval windows. One of these rooms, clean and luminous facing the main court, stirs Franco's imagination:

"Oh, if I only could make my laboratory here!"

Franca turns to him:

"You certainly can! We can do whatever we want here! ….and this one was already used as a "home" laboratory by Grandpa Emilio, who owned a car paint factory".

Franco shouts: "WOW! A lab, to a scientist is like a bike to a cyclist. And without a lab I will never know how much I really want to become a researcher …"

The girl is thinking ...then says:

"I never thought about that! How did science come to your mind?"

"Well, my Dad is not a chemist, but says he always liked chemistry….and last year he gave me a booklet about the elements…so I know about the *Acids* and the *Bases*…. It all depends where Elements are written in their Table…. then there was a movie, the last I saw before we found ourselves at war also against America."

"You don't mean *Tom Edison*!? I saw it two years ago: that boy was mad! Every idea he had was a disaster!"

"Franca, hold your horses! Movies are never all true, you know that".

"A lab could be handy for me as well in case I decide to become a Biologist of some sort! This room has not been used for a long time and I bet it still contains some equipment".

"In any case, I need a helper to make a lab, decide the experiments and carry them out. Would you like to be my assistant?"

Franca does not need words. Her eyes cannot hide the enthusiasm. They shake hands, and she says:

"And I will get permission to use the room, do not worry, but what experiments are we going to do? Do you already know? Can you tell your assistant"?

Franco is elated but he does not give a precise answer:

"It is wonderful to have a lab! But to experiment we need something to experiment on: a substance, a plant, insects, salts, acids. We cannot buy these matters… but I am sure that if we continue our exploration of this building we will find many substances and also ideas of what to do with them!"

"Yes, I like that! I will take you to the cellars, there's lots of stuff there! Are you afraid of mice, by the way?".

Before Franco can utter a reply, she gives him her explosive laugh.

The cellars and the storing rooms for gear and produce from the vegetable garden are also majestic and Franco and Franca find useful things for the lab. From a large basket they pick some Spanish beans (white, large beans). Franco explains:

"These are for the germinations".

"How many will we need"?

"At least twenty-five, considering controls".

From another basket further down Franca throws an apple to Franco, and bites one herself with gusto. In a darker room on the north side of the building they make a discovery: the humid wall seems to secrete a whitefish fluffy kind of snow…Franco shouts:

"Look at this stuff! I think I know what it is: SALTPETER! A very important salt. We must collect it, and we will purify it in the lab! Hurray!!

While they are scraping the wall with a chip of shingle and filling a box with the alleged Saltpeter, the cantankerous custodian Natalina appears, made suspicious by seeing light in the cellar. She looks at the youngsters:

"What the hell are you doing here!? Franca, does your father know?".

Franca and Franco have a hard task explaining their actions. Communication must cross too many barriers (dialect, generation, social class, country/city gap, etc.). Fortunately, Natalina gives up and makes her grand exit.

"Where is this world going?", she says, and, lifting her fist to the sky, launches her signature warning:

"A' DA VENI' BAFFONE!" (Translation: "Wait until Big Mustache comes to set things right"! Referring to Stalin).

On the way to the stairs with the new goods Franca spots some bottles on a shelf. One is half full and she reads the label:

"K.E.R.O..S.E...". She makes a grimace…" What is this smelly thing?"

"Something like petroleum, gasoline…brownish, dirty… we can use it"!

In the laboratory, after distributing the beans (three per dish) in a series of dishes containing cotton, sand and water, to which they plan to add different substances and monitor their individual effects, Franco changes his mind about the kerosene:

"We will use it, sooner or later, but it is flammable, and this lab is mostly made of wood…I prefer to take it to a safer place, near our bathroom….at the bottom of the tower, there is a little storage room" …

Franca looks at the boy and thinks: "How thoughtful of him …"

To Franco, Sfollamento in Velate becomes more and more exciting, even though most of his time, Monday through Saturday, is occupied by school and the commuting to school. The trolley line is the same line as last year, but the travel time is longer. It takes half an hour to cover the five Kilometers that separate Velate from Varese.

A dozen students must leave home at 8 a.m. sharp and march along a narrow street in the woods to reach the Trolley stop. The street goes steep down, crosses a bridge over a stream and goes winding up to the trolley station. Most of the students try to arrive some minutes in advance to be safe.

However, Franco and Carlo Vandoni belong to the few who hate to start too early, running all the way for fear of being late, and facing the possibility of waiting for a later trolley. They begin to gather data.

Luckily, before the street reaches the bridge, there is a vantage point from where the trolley, which has left *First Chapel* station, heading down to Velate becomes visible for a few seconds in a sharp curve and emits a resounding screeching sound before disappearing behind tall pine trees.

By timing the interval between *screech* and *boarding* at Velate Station for several days, in different weather conditions, the boys determine that there are three possible cases, each suggesting one satisfactory behavior:

1. *If one hears the screech before reaching the bridge, the trolley is lost and he will walk leisurely to the station and take the next trolley in about half an hour.*

2. *If one hears the screech while he is on the bridge, he will have to run like crazy in order to board the trolley.*

3. *If one hears the screech after having crossed the bridge, he will catch the trolley without having to run.*

These calculations turn the monotonous morning walks into an exciting mathematical game. As per an unspoken agreement, on board of the trolley and during intervals between lessons, every boy will disclose the decision made in the morning and seek approval from the group, creating a full-fledged social competition.

The school this year is the local family run establishment *Istituto Beccaria*. Director and Assistant Director are man and wife. He teaches Italian. She teaches Mathematics. The rest of the teachers are either young and unqualified or qualified but too old, like Prof. Rodolfi, who wears thick glasses and is almost deaf. All of them are probably underpaid. The setup seems blueprinted from a popular book of the Thirties, *Gianburrasca* (Hale-Storm Johnny).

The Beccaria has both "boarders" and "non-boarders" and the first disappointment for Franco is to find out that his friend Carlo is among the boarders and unavailable to play with after school, except on Sundays.

Santo Stefano is the parish church of Velate, with red and white ornate stucco, and a tall bell tower dominating the hill. The energetic parson (Don Enrico Alberio) has seen his flock and his engagements suddenly burst and is thrilled about having his tiny parish complemented by so many metropolitan newcomers, mostly middle-classers from the Big City of Milano. He responds positively to these changes but knows that the principal factor of revival is the war itself, with the presence of danger, the need to talk to peoples and to be reassured and to be told that there is hope, the hope of a world better than the present one…

The original population of Velate must have been around five thousand, the parish counts a larger number of souls, because tiny villages around, like Avigno are historically part of Velate and in care of Don Enrico, together with a few other Chapels and little churches in the territory. The parson is quite busy, considering all the trips to see sick and needy people, saying Mass in all suitable locations, teaching Catechism to the children and organizing soccer matches in the Oratory. Luckily, he is a proficient cyclist and seeing him speeding up or downhill around the territory is a frequent occurrence, his black pedaling shoes, black stockings and black knickerbockers revealed by the tucking up the frock at his belt.

"I bet I can ride around with my eyes shut or by night with no lampposts?"

He almost wins the bet in one occasion: April 1,1944, the night when the RAF first attempts to hit the Aermacchi factory and fails, but, in spite, drops a few bombs around Avigno (one of them falls 500 meters from Velate as the crow flies)[100]. It is midnight, all Sfollati hear the explosions. Don Enrico climbs on the bell tower, sees the sky illuminated by rockets and some fires behind the trees. The planes make a last circle, their roar fades and they are gone. Then the long sound of End of Danger. Don Enrico is worried. Phone lines are off. He must go and find out what happened to his parishioners of Avigno and whether they are in need. He will leave at dawn. At 3 a.m. it is still pitch dark, but Don Enrico cannot wait anymore. He

---

100. The one on Sunday April 30, is the second attempt to hit the Aermacchi in 1944.The first attempt took place on April 1, by night, with great display of fireworks to illuminate the scene. However the RAF targeted the wrong valley: that of Avigno instead of that of Masnago, one mile away. There were no victims. Few bombs were discharged; one of them probably defective, exploded with a 12 hour delay, which scared and puzzled the Sfollati at lunchtime.

jumps on his bike, pedals in gentle downhill to the Velate Tower, a Roman relic now illuminated by the fading moon. He easily finds the connection to the more severe downhill winding road to Avigno. His wheels seem to remember the right position for every curve and Don Enrico knows exactly where he is: in a couple of seconds he will transit on the left bank of the road and make the connection to the flat, wide central street of the village. Except, there is no bank, and even no street: only a crater created by the last bomb dropped; Don Enrico is the first to see the ghastly hole as he dives into it with his bike, and both are hurt.

Fig.55 - Don Enrico falls into a bomb crater

In no time the Avignoese are alerted and practically all of them come to the accident's scene and pick up their Parson, put bandages around his shoulder, clean his wounds, look at him deep in the eyes.

They had not expected this extraordinary act of caring for them that this priest, energetic, but of few words, has shown. They think: "If it takes a war to experience such an act of love, we have a lot to learn about this world". Two youngsters, improvised mechanics, go pick up the Parson's bike: it is a mess, but they will fix it like new!

The church choir is all-female, composed exclusively of local residents.

Fig.56 - Velate's church choir

Their voices are marked by the dialect of the region, even less comprehensible than the Milanese, with vowels somehow brutalized (or "thrown away") at their emission, a legacy of the northern barbarians, who came to pick up the spoils of the Roman Empire. Besides the 11 o'clock Solemn Mass of Sundays and festivities, the choir, led by the Don, also sings the daily Vespers at 5 pm. All songs in the repertoire are in Latin. The most popular one is *Ave Maris Stella* ("Hail, O Star of the Sea"), a song of praise and prayers to the Virgin Mary, dating back to the High Middle Ages, but the choir has problems with all this Latin, they do not understand the words and feel the frustration. They want to put their heart in this prayer: "Especially now, with the dangers of the war, and the bombs being dropped, we need Mary's help more than ever. Will she listen if we don't understand what we are singing?"

They start strong with the first verse, but then continue sheepishly with the nonsense. The daylight decreases, the bystanders Sfollati are condescending… Suddenly the hymn comes to life, loud and clear: "MA LA NOSTRA PELLE!" becomes a real invocation, very comprehensible. It is sung with such emotion, it cannot be ignored! Carlo and Franco (in their second year of Latin) look at each other and understand that the choir has read "*Ma la nostra pelle*, ("But our own skin") as a pressing request to Mary (in Italian) to save their lives, instead of the original MALA NOSTRA PELLE, (i.e. "Dispel our Evils"). The boys think the choir has made the right choice.

Sfollati have a network of communications based on a few phones located in grocery stores, other private phones and word of mouth between the women of the families, most of which knew each other in Milano. The network covers a dozen villages.

Through this network, on a perfect-weather-Sunday of April 30, Franco is invited by the Longoni, to a hike to the top of Mount Martica, plump, 900-meter tall and all covered with grass and flowers, which dominates the view of Lake Varese on the south and of the hills approaching the Sacro Monte on the North.

Stefania's father, Alberto, indicates all the villages (Velate sits at the foot of Sacro Monte) to the hikers who begin to open their sandwich bags: a lovely excursion. Looking straight down to the base of Martica, Franco sees a shallow valley, containing a group of orderly industrial constructions, with several smokestacks, well visible despite the disguising colors.

"What is this…?". Before Franco can pronounce his question, Alberto continues:

"We are facing Masnago, and the plant you see straight in front of us is *Aer-Macchi*, where the fastest Italian planes are built."

Franco knows a lot about military planes because a family friend (Air force Major Tagliabue) used to give him catalogues of British, German and Italian planes with secret data about their weaponry "strictly reserved to the pilots" and is a fan of the new *Macchi-Castoldi 202 Folgore* ("Thunderbolt"). In fact, almost every day, one of these fighters, surely better looking than the British, makes an appearance in the sky with gracious evolutions in all directions and ending with a deep, long dive followed by a majestic rise. Franco will not forget the modulated sound that this creates.

The hikers (a party of ten or twelve) are seated on large granite boulders emerging from the grass and are half through their sandwiches when they all stop chewing: something from the right end of the horizon is moving to the center, growing rapidly in size and producing a roaring sound. The objects are now recognizable as planes. Alberto has extracted the binoculars from its leather case and monitors the situation:

"Bombers! Probably British! Many, this time in mid-day[101]"! He is trying to count: "8, 9..12...15!" then, he shouts:

"They are coming towards us!! Everybody lie down! Try to hide behind those rocks. Don't move!"

Now the bombers are visible without binoculars; they are in a geometrical formation of three by five. They have reached the center point in front of us, higher but not so much higher than the mountain, and they start a half-left turn, which brings them directly towards us. The turn is signaled by the glittering cabin windows reflecting the sun rays change of angle and by the sudden rise of the roar's volume, as if announcing the show. Little clusters of black dots appear under their bellies; the planes begin another left turn.

101.When the hikers saw the planes from the vantage point of Mount Martica, where were the Velate friends and family? There are precise testimonies from Carlo Vandoni, 12, and Gianni Celada, 9. They were exiting the Velate church with their families after Sunday mass when they heard the awful sound of engines and saw the circling planes. As they saw the bombs dropping from the planes, they had the terrifying impression that this was happening directly over their heads and froze in scare. Comparing memories seventy-some years after the facts, it is extraordinary that the same crucial event captured from two different points of view, yielded a sort of 3D picture, totally coincident and corroborated. As for the sense of imminent danger experienced by Carlo, Gianni and the other church goers, it was fortunately not justified by the aerodynamical forces involved: when the RAF fliers dropped their payload at the Zenit of the Church of Velate, they were in fact aiming at the Macchi plant 4 KM away. The bombs, released at the height of 1000m, pushed by the carrier's speed (360Km/h or 100m/sec), pulled by the Earth's gravity and slowed by air resistance, would travel for an entire minute before touching ground at the foot of Martica. Thus, the nearest explosion to Velate happened, at least, 3 KM away.

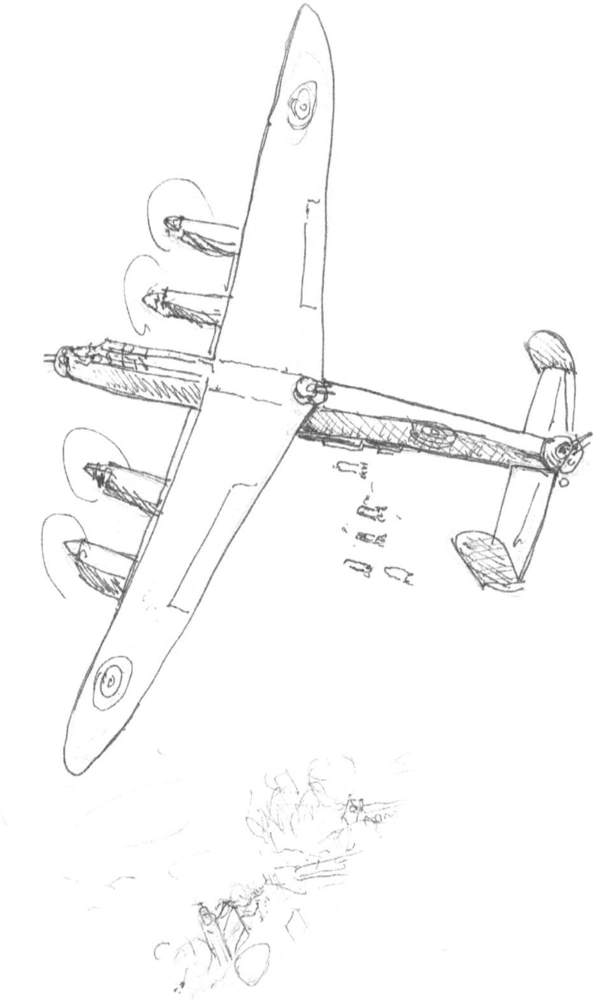

Fig.57 - Lancaster bombers attack the Aermacchi Plant, 1944

Fig.58 – Velate: church goers see bombers over their heads

The bystanders' eyes are staring at the plant below, which seems untouched. Very long seconds pass. Growing puffs of smoke start to germinate in regular rows from the factory. A silent show, which is followed with a delay of three or four seconds by a horrific thunder. The hikers are silent, their eyes' eagerness to see more is defeated.[102]

The site of the Aermacchi Plant is occupied by the combination of explosion smoke and dust of crumbling structures, both inexorably growing. No one can tell whether the plant still exists or has been substituted by the opposite of structure: the growing entropy and chaos.

When the hikers begin their descent, the basket shaped valley still keeps its secret, filled by a homogeneous impenetrable cloud. Alberto is mumbling:

"*Vigliacchi!* ("Cowards!") They hit us when our army is disbanded!"

Franco asks Stefania:

"Where are the MC-202? Why did they not show up in defense of their mother-plant?"

Stefania surmises:

"Perhaps because the MC-202 is only one, flown over and over every day…if so, it was wise for its pilot not to attack 15 bombers." Franco agrees, remembering the awesome firing turrets on the Lankaster shown in the Secret Intelligence booklets. There is no time

---

102. For the daytime bombing of April 30, RAF must have provided better maps and the target was hit and destroyed. Being Sunday, there were no victims at the Macchi site. However, a Military hospital had been foolishly accommodated right behind the factory in the former *Grand Hotel Kursaal*. It was clearly labeled with a large red cross: the building was not hit, but several bombs fell in the surrounding garden. By a terrible twist of fate, many of the recovering soldier patients had the instinct of running to the garden, as they would do in case of Earthquake...many lost their lives.

There is another episode regarding the same bombing: a stray bomb hit a stable belonging to the *Villa La Quiete*, a Private Clinic then owned by cousins of the Franchini. There, Fate was in a much better mood: the bomb killed two pigs, which were bred for food and were named *Benito* and *Adolfo*.

for further conversation: the hikers are running breathlessly downhill, carrying the heavy load of the drama they have witnessed, and hurrying to reassure the families about their own wellbeing.

In the Lab there are two lines of experiments going on. One is the quest for an artificial soil, esthetically pleasing, with the energetic germinations protruding from the beans. But Franco realizes that, while they can easily disturb or stop the germination by adding a number of substances, such as salts, or acids, if they try to favor the plant growth, none of the additions tried result in better growth than pure water. He is puzzled: how come that a growing stem does not need any added food? It is impossible, it cannot be true! Franco and Franca engage in a long discussion, and conclude that seeds are not the right object to test to determine what addition may improve their growth, because seeds (and beans are seeds) are a natural package of all dry stuff necessary to the germ until the new plant becomes able to suck up nutrients by its own means. Asking what other nutrients a bean needs is like asking whether an egg must be fed to produce a chick: it is an ill posed question. The instructions are: Just add water, the jumpstart of life.

The second line of research is more promising. Using some beakers and bottles forgotten in the lab, Franco launches the "purification" process. He fills a beaker with the shaving collected from the wall in the cellar, then pours water slowly while mixing the contents with the help of a wooden spoon, and obtains a thick suspension, partly granular, grayish; he adds more water, mixes again and manages to separate most of the solution. He discards the solids by transferring the cleanest part of the concoction to another beaker. The second beaker is then filtered through a thick cloth that Franca keeps stretched over the mouth of beaker n.3 by pulling it with two hands. The filtered liquid is almost limpid and almost colorless. Franco does not resist the temptation of wetting a finger in the solution and touching the tip of his tongue. His eyes shrink. Franca asks:

"What does it taste like?"

"Saltpeter! What else?" Then he meditates: "Saltpeter is one of three components of Black Powder..."

"We found an explosive? In our own cellar?"!

"No! It becomes explosive only when it is mixed with the other two components: Carbon and Sulfur. But we are not going to do that mixture."

"I see, and I also see that our non-explosive is dissolved in water"!

"Correct! We have to bring it back to the dry state. The best way is to crystalize it…it will take time. This solution is very concentrated, and if we leave it in a shallow dish in front of the ventilator it will dry faster."

The two set up the dish in position, and also a second dish with stripes of blotting paper, after having immerged them in the concentrated solution.

Daily life: the Sfollati network has located an English teacher three kilometers southwest of Velate and several boys begin to attend private lessons. Franco will take his bike, a Bianchi 20" with wooden wheel rims and attend once a week. He is happy to keep up the English he has learnt during his last year of German school. But also German will be kept up: Alice tells Mimma of a lady from Leipzig, who cannot get a regular job because of the Racial Laws[103]. Frau

103. Memories related to Nazism, originated in the years1936-41 (when the author was 5 to10 years old), are contained in the chapter *Stefania*. In 1938 the Racial Laws enforced in Italy signaled Mussolini's definitive domination by Hitler

Hirsch will give lessons to the Celada, the Franchini, and the neighboring Vandoni and Quadri.

The streets of Varese are patrolled by the occupants, the *Wehrmacht*. The presence of the Nazi SS is not immediately obvious. During spring 1944, the *Repubblichini*[104], the various corps of the fascist Militia, are seen more and more. They wear new uniforms with knickerbocker pantaloons touching their shoes, and pose either as *Brigate Nere* ("Black Brigades") or *Maro'* (short for Marines) of the *Decima Mas* ("Tenth Squad of Motorboats Anti- Submarines").

One morning, the students headed to the Istituto Beccaria see four Maro', only few years older than middle school students, intent on boarding a trolley in motion. They are standing on the outer steps and showing off the pineapple hand grenades at their waist, their automatic weapons under their shoulder, *Machinepistoles*, courtesy of the Führer.

More surprised than horrified, Franco's classmates comment this exhibition at after-school, where they are supposed to do their homework. The supervisor, young extrovert Miss Pandoro, intervenes:

---

and looming disaster: a slow process with no glory for the Duce but with much sufferance for Italians. Hitler, perverse and mad, trying to distort Reason and Nature, is doomed. The Church survives by the invisible resistance of millions of faithful rather than by the ambiguities of Vatican City. Pope Pius XI, had signed Concordats with Italy in 1929 (calling Mussolini the Man of Providence) and with Germany in 1934; when he decided to "come about" he wrote *Mit Brennender Sorge* ("With Burning Concern") in 1937, refused to meet the two dictators in Rome and planned the Encyclical *Humanis Generis Unitas* condemning racism, but died in early 1938, before publishing it. His successor and former Secretary of State, Pius XII, allegedly afraid of consequences for the Catholics in Germany, buried the drafted document in the archives and opted for silence. The draft was discovered after 30 years and published after 50.

104. All soldiers, Brigadiers, Marines, Officers, politicians of the *Repubblica Sociale Italiana*.

"Yes, they should be more prudent, especially with those grenades", she goes on, dreamy..."I met a two of those boys last week….they are so full of life, so sweet and idealistic!"

Fig.59 - Maro' terrorize the town riding on the tram

These remarks conclude the argument, but the teacher's new friendliness induces Mastelli to bring up something completely different.

"Miss, I speak for the class: do you know that this morning we had a Latin test and it went so badly that we are all desperate?

Miss Pandoro is surprised. She is surrounded by mourning faces:

"Do you already have the results? I am sure Prof. Rodolfi didn't mean to assign an impossible task: you must be exaggerating!"

"No, no! It was terrible, he gave us a text by Lucretius. I wish you could see it!

"As a matter of fact, the professor takes his time before correcting, and he keeps the tests in that closet by the wall..."

She walks to the closet, easily opens the door, picks up the pile of translations and returns to her desk...

Fig. 60 - Miss. Pandoro corrects the Latin tests

"I will be able to show you that your fears are unfounded!"

The class cannot believe she was so easily convinced and stares at her. Pandoro looks at the text and the translations. She makes a grimace. She looks again, then:

"Actually, the text is not simple. Old Prof. Rodolfi sometimes...." She shakes her head.... "and, from the few I have seen, your translations are terrible ..."

More whining erupts, this time vocal, as in a Greek tragedy. Pandoro thinks some more, then:

"I am sorry, boys. The only thing I can do is show you where you failed and this will be a useful lesson. You can take back your text; I will translate aloud….

But before she begins something clicks in her head:

"But remember, you are responsible to put it back where it was, because I have an errand and must leave early today".

The majority of the students rush to take back the tests. Franco is among the few who don't, but has to withstand a surreal insistence delivered by voice and muscles:

"C'mon! Join the majority"! "Don't be a fool"! "Don't ruin your school rating"! "Don't let us down"! "Don't be a damn traitor. Think of us"!

In the lab, Franca and Franco evaluate their progress. In the Saltpeter line they see dry crystals on the blotting paper strips, while the free solution is still wet. Approaching one of the strips with a lit match. The strip does not catch fire, but they hear some sizzling, and some local eruption of the flame.

Franca: "Why does it not burn better?"

Franco: "To burn, Carbon (C) must combine with Oxygen (O), and make CO2. We have O from the Saltpeter, but not enough C in the strip…. Wait a minute! Does Sugar burn?"

"Yes...it happened last week, when Mom was making caramel cake, and forgot the pan on the fire. The sugar became brown, melted, then it became carbon, and then there were flames and smoke…"

"And then"?

"Mom had to throw away the pan".

"You gave me an idea! We can make a soluble explosive. We have the Saltpeter solution. We must find the sugar and dissolve it into the Saltpeter in the beaker. That's it!

"I can steal some sugar from the kitchen… but not much, you know that it is rationed! But why do you want to make this soluble explosive? To blow up submarines? To grill fish before you catch them"?

"Stop laughing, Franca! Obviously, when dissolved in water our stuff does not explode! Anyway, the crazier the idea, the bigger the discovery, if you can find a purpose for the experiment. Don't you agree?"

Franca does not answer. She exits the lab, to go scouting for sugar.

Prof. Rodolfi finishes the correction of the test and he is puzzled. His first reaction is:

"My assignment was too easy, next time I will give them something more engaging …but he is still puzzled: some of these students are real morons. How could they improve so quickly? And why were the critical points identical in so many tests?"

Franco gets a 6 minus (passing) on his test, and all considered, is happy. The day of the test Mimma had asked him "how did it go?", and he had told her about the afterschool facts, only after she promised to keep it secret. Six days later, Middle Class 2 is notified that the recent Latin-Italian translation is canceled and a new test will take place in a week. Franco and the other dissidents have to endure the unpleasant feeling of being called spies by their righteous companions.

In the Lab, the blotting paper immersed in the saturated solution of Saltpeter and Sugar is now dry and can be tested.

"It tastes so much better than Saltpeter alone"!, approves Franco.

The charged blotting paper is stuffed into a piece of copper tubing eight inches long protruding from both ends.

"Franca, strike a match and light the tubing at your end!"

"Done as specified, Boss!"

Reassuringly, this operation looks more like lighting a cigarette than a firework. There is no flame. The fire advances, sizzling and smoking, penetrates into the tubing and disappears. Four eyes are eagerly watching……after a couple of minutes the fire emerges at the other end.

"Hurray! We have invented the *Soluble Fuse*"!

And the experimenters engage in a wild dance around the room.

Fig. 61 - Franco and Franca celebrate

Easter holiday brings back Carlo Vandoni from his 'prison' and automatically recomposes a well-rehearsed hiking team. This time Franco and Carlo are set to explore a thick forest at the foot of *Campo dei Fiori*, the highest mountain around.

The walk is longer than predicted and expected by the moms, Adele, and Mimma. The boys know that they will have to walk fast, in order to be back before dusk.

The vegetation is dense and, instead of a regular path, the boys follow what seem to be animal tracks. When they reach an unexpected opening, they get a good view and spot two trees covered with tiny purple fruits. Carlo shouts:

"This is the Wild Cherry spot!".

He puts a handful of cherries in his mouth:

"these are the best cherries I ever had, and, probably, will ever have in my life!"

After feasting, the boys go back to the opening that led them to the cherries, an area with stratified, partly broken rock and no vegetation. They find that the fissures in the Schists contain a black, humid paste. Franco puts his finger in the stuff, then to his nose…:

"You know, Carlo, it is very similar to something some Friars put on my foot in Sestri Levante, before the war! While swimming, I hit a sea urchin, and my heel was full of spines…."

"Wow! Did it work?"

"Well, the next morning the spines came off."

"Then this might be *Ichthyol*, or similar. A product of pre-historical life, like mineral oil. What a discovery! Are you ready to go back home?"

"No, not before I collect a good sample for the lab!"

Two weeks later the local news mentioned an accidental shooting in the Maro's barracks. One soldier, 18, perhaps one of those Franco saw boarding the trolley, is dead. A big funeral is announced. The newspaper insists it was a chance misfiring, however a party ending in a brawl, jealousy, girls, alcohol, are suspected. The class wonders if and how their teacher will be affected...

The next afternoon the door of the afterschool is closed, with a handwritten note: "Afterschool temporarily cancelled. Signed: Vice Director". Is Miss Pandoro sick? Nobody asks. Nobody tells.

The class senses there is a connection with the shooting in the barracks and the image of the sweet, scatterbrained girl takes on a tragic shade. When she reappears, two weeks later, it will be for the last time. Her eyes swollen, she is a broken person. She speaks softly, tentatively:

"I am here to say goodbye … I liked you boys. I wish you a better life than I have been given… please remember me! My name is Lisa" and she is gone.

Exploring the stable, now hosting one single cow, to provide milk for the owner's family, Franco picks up a light green flask. It looks a bit like the Chianti wine fiasco stripped of the straw protection, but smaller. The glass is perfect, very thin and Franco has heard that blown glass is used in all laboratories because it resists fire. He thinks that he has found the missing link that will enable him to experiment distillation. He is fascinated by this possibility, and talks to himself:

"One takes a solution containing many liquids, or one liquid and any number of other dissolved substances: if you boil these solutions, the liquids become vapor, while the salts remain in the flask. The vapors, if taken away from the heat, change status again becoming liquid and *pure*".

Franco is not interested in trying to distill wine because it is not available and also because everybody knows the result: Brandy, i.e., concentrated Alcohol. Instead, who knows what may result from distilling ...that dirty Kerosene? "I have the flask, the rubber tubing, the Bunsen Flame... I can do it, therefore I must..."

What Franco does not know is that by this kind of reasoning he risks losing the right to use his real name and go back to being Francolino, and even much more! However, his subconscious tries to tell him something: he will try the distillation alone and not in the lab, where the risk of starting a fire is bigger than in the stonewalled Celada kitchen, without telling Franca as she would pose too many reasonable questions.

Today, Mimma is out on a mission to find some meat in the black market, Guido is in Milano until Saturday and Gianni is invited to play with Susi.

Franco has set up the materials on the sturdy kitchen table. The flask, held by the neck with a clamp, and containing about 100 cc of Kerosene, is positioned over the Bunsen burner, without any protective iron net. (Without Net?! Why? Because he did not find the net). A red rubber tubing connects the neck of the flask with a large beaker on the table, immersed in water to favor the condensation of vapors.

This is probably one of the fastest experiments ever conducted (less than 5 minutes). Franco lights up the Bunsen flame, blueish/orange, which elegantly embraces the bottom of the flask. Some bubbles begin to appear, as expected. Not expected is the strange, nauseating smell of rotten-eggs, that fills the room, caused by the hot Kerosene flowing through the tube and melting the rubber. Franco becomes lightheaded.... Suddenly a dry, metallic, acute sound goes off and, at the same time, he sees a hair-like curving crack developing on the flask and becoming decorated by a tentative line flame on the surface of the glass. Franco has the reaction of a squirrel: he turns off the gas and dashes out of the door that opens onto the garden. He waits there for ten minutes, not believing his luck (or was it a guardian Angel?). When he re-enters the kitchen, it's as if time stood still and the ready-to-explode flask has stopped one second into the breaking process and the spilling of its contents of havoc

Fig. 62 - Franco escapes a near explosion

14

Sfollati - Part Three

Half-way between the Manor's gate and the tower, and four meters into the large lawn, called *Il Pratone,* is where a swarm of hornets has decided to settle. The Italian name for these giant yellow striped wasps is *Calabroni.*

Fig. 63 - Cross section of a subterranean Yellowjacket nest showing the various tiers with cells.

Up to two inches long, Calabroni are dangerous, and this particular colony must be large, judging from the traffic in and out of a black hole in the lawn, the only visible spot of their underground nest.

The mission is to get rid of the intruders and all the kids of the Manor, except the girls, participate. A long bamboo cane is found in the stables, and Franco comes out with a new purpose for the rest of the ghastly Kerosene.[105] He empties the bottle in a tin, secures the tin to the tip of the cane, holds the cane like a fishing rod and, when it will reach over the nest, plans to rotate the cane on its axis like a giant ladle, thus spilling the Kerosene onto the objective.

If they make it that far, the boys will also try to light a fire, but this part of the matter has not been developed yet. The cane is thick, long, heavy and it needs three boys to bring the payload in the vicinities of the target, while keeping a safe distance from it. After some trials, the tin seems to be in position.

"Ready, set, go"! Commands Franco.

Six hands begin the rotation… when two roaring flying objects hit the "commander" on the forehead and Luca on the nose with the force of a punch.

---

105. The same bottle of dirty Kerosene, which almost caused an explosion in Franco's kitchen a year ago, is still not empty…

Fig. 64 - The revenge of the Hornets

All boys let go of the cane and get the hell out running. Only when he reaches the arched court Franco slows down. The girls are playing prisoner ball; Franca shouts:

"We were waiting for you! Here's the ball!

To catch the ball Franco must move one step to the right, but his knees start to wobble and he collapses. The boy, on the floor and unable to stand up is surprised, and understands that this is the "*Revenge of the Hornets*". When he is escorted home and put to bed, his face is full of tiny red pimples and his breath is heavy, even if the site of the sting is not swollen. Mimma and Giulia are around and the latter calls Dr. Taborelli, the physician in charge of the district extending from Varese to Sacro Monte. He has seen his patients and responsibilities increase exponentially with the arrival of the Sfollati. Dr. Taborelli arrives in twenty minutes driving his car, recognizable thanks to the little Red Cross flag on the hood, which makes it the most popular of the few vehicles allowed on the streets.

Fig. 65 - Dr. Taborelli runs to Franco's rescue

A short visit is sufficient for a diagnosis:

"This is a type of allergy, called *Acute Anaphylactic Shock.* I will tell you about it, but the case is serious and is still developing: I have to start a therapy right away, before the swelling of the glottis impairs breathing".

"Tab!" breaks in Franca's dad, a longtime friend of the Doctor "Are you going to use those new *anti-histamine* drugs….?"

With an unhappy expression, Taborelli says:

"I wish I could, Fabio, but our supplies are incomplete and do not include new drugs…. I will probably inject adrenaline, with extra care.

Fabio replies: "Look, I got this in Lugano last year for my hay fever, and there are two vials left. I will be glad if you can use them on our wasp-hunter here!"

After the first injection, the other boy stung by the Calabrone comes to see Franco. Trying to smile, Luca Franchini, six-year-old, stands at the bedside, figuring out what to say. He has been hit in the middle of the face, in the soft tissue on the bridge of the nose. He is unrecognizably blown up. His eyes are replaced by narrow slits; blond and swollen he looks like a little pig.

Mimma asks the obvious question:

"How come these two boys, bitten by identical insects, have such different symptoms and Franco seems to be so much sicker?"

Dr. Taborelli appreciates good questions and says:

"Dear Madam, your observation is logical, but today's medicine may not have a satisfactory answer. I will try to approach it forming a few hypotheses and seeing how they sail. In the case of Luca, who is younger and had fewer occasions of experiencing wasp bites, the symptoms reflect what a poison does when it is injected through the skin; It causes damage to tissues, which in turn produces *Inflammation* and this time the swelling is so enormous because Calabroni compared to regular wasps, carry a large dose of poison…his Immune Response is *Primary* and takes time to develop: the boy will feel better in about a week and his blue eyes will become visible again. Franco's case is different: I suspect his sting was not the first for him".

Mimma whispers in Giulia's ear: "Three years ago!"

The Doctor overhears, nods and continues:

"At the first encounter he developed his Primary response. As I was saying the Primary is slow, but a most important discovery tells us that copies of the primary cells are left behind and stored in the body for decades in a "Ready to go" state, just in case the same "antigen"[106] shows up again. If so, the secondary response is triggered and will act in minutes! We call this mechanism *Specific Memory*. Memory makes us prepared to respond, faster and better and is essential in preventing infectious diseases. We learned of this kind of Memory first from Jenner, a country doctor in Great Britain, and later Pasteur, in Paris, and thanks to them *Vaccination* has prevented most of the worst infections.

What happened to Franco is a full-fledged "secondary" response, where the antigen injected by the wasp acts as trigger of a general Immune reaction and the response itself is causing the discomfort, including the danger of suffocation: a self-inflicted ailment or, in military terms, a *friendly fire!*".

Mimma is utterly confused:

"Why is the vaccine a good thing and the secondary response to the wasp bite dangerous? Is something wrong with my boy?"

Dr. Taborelli thinks: "I must be careful with these moms!" And smiling:

106. See *Antigen* definition, footnote n.4, p.31.

"No, no, nothing is wrong. The problem is that Immunology is a young science and we are still very ignorant. What we know is that a group of immune responses, including those caused by insect bites, poison ivy, jellyfish stings and flying pollens, are incredibly disproportionate compared to the danger they are supposed to fight. All scientists have done so far is to group these cases and to label them "Allergies", which simply means "different responses" .... I am sure this field - so important - will soon be developed by specialists. For the moment, you must do with the experience of your country doctor...... However, I am sure there must be some sense in allergy, since evolution allows it to exist. I wish we could justify these reactions and maybe learn to *use* some of their energy instead of only trying to block them ...with drugs full of side-effects.

Franco seems to be getting great relief from the anti-histamine and, after listening to the doctor's speech, cannot restrain himself from declaring:

"One thing's for sure: from now on, I'm staying the hell AWAY from wasps and giant wasps"!"

"Bravo! "Comments the doctor, reaching the boy's hair and ruffling it with his fingers; then continues, as if thinking aloud:

"You have been scared by the anaphylactic shock! And this modified your behavior and that of your bystander friends and made it *safer*! You are giving me an idea: the shock is not a cure; it is a *message!* This effect may be *life preserving.* So far, I was unable to imagine a positive feature to justify allergy in any of its shapes.... but in today's conversation we are stumbling on a very interesting lead! Could the selective advantage for the shock reside in an educational, *warning function*? ultimately in *Information*?" Taborelli is now unstoppable:

"I am thinking of animals or primitive men, with limited communication with each other... If something violent and impressive happens after contact between an insect and an individual, the bystanders will be warned and possibly saved by becoming more careful: a social evolutionary benefit of immunological memory! And an advantage in the competition for existence".

September 1944: another change of school for the two eighth grade classes of Velate. Both have reasons to welcome a change of school, but those of Carlo are stronger, and, in part, contradictory. He will remember the year at Beccaria as the worst of his entire school life: he felt "imprisoned" in the boarding school routine against his will; he suffered the pangs of hunger since the food was bad and scarce, an art teacher was picking on him. On the other dish of the balance, is Lia: one of the few of girl's boarders. She is an offspring of the Rame Family of actors, touring from village to village and performing off a Thespian Wagon. Lia's younger sister, Franca Rame, will marry Dario Fo[107], and create with him a world famous Satirical-Radical Theater. Carlo had a crush on Lia. The first time he saw her she was arranging the chairs in front of the stage. He remembers her as handsome, taller than him and just her being at school helped him stand the stresses of Beccaria.

The San Carlo will certainly be a better school. It was one of the solid traditional Catholic Schools of Milano, which has lost most of its students (and their fees) because of the Sfollamento and was temporarily transferred to Varese in order to get them back. The facilities are precarious, but the teachers are good. Don Mandrini, a frail looking priest who runs the Seventh Grade class (Terza Media) shocks the boys on the first day, when he opens and begins reading

107. More information on Dario Fo and Franca Rame in Sweden in Chapter 8, *Gilbert*.

La Roba ("Stuff")[108], a dramatic meditation on human greed, by the Sicilian author Giovanni Verga. Franco will remember the forceful rendering of the socially charged prose and the swollen vein in the priest's neck as he read the last sentence at the top of his voice:

> *"When they came to tell him the time had come to leave his Stuff and mind his soul, he ran out to the courtyard like a madman, faltering, brandishing a stick, killing his ducks and his turkeys, shouting: "You are My Stuff, you are Mine! You come with me!"*

This is going to be the best year of his entire school life, and probably also for the majority of the students, including Carlo, despite the fact that San Carlo, being a Catholic Institute, is many decades away from becoming co-educational. After learning that the Don is not a frail character, the second surprise of the 1944-45 Schoolyear is in the realm of Physical Education.

There is no gym on the San Carlo premises, and the Class walks to a rather far, unheated industrial construction illuminated by skylights where a tall athletic looking man in overalls orders the boys to run around the empty room a number of times and do breathing exercises. Then, he orders them to stand on attention in line facing him, and speaks:

"*Ri-poso!* ("At Ease!") My name is Furio Gallini: like all Gymnastic teachers I am in charge of collecting your voluntary enrollments to the New *Opera Balilla* for the *Repubblica Sociale Italiana*. You probably were Balilla before…. The New Balilla wear a modern uniform, and

---

108. *La Roba* was first published in a single volume in 1883 and later collected short stories *Novelle Rusticane* ("Short Novels of Sicily") in 1882.

represent the Republican and Social values.... Those who wish to enroll: Two steps forward: ONE-TWO"!

Gallini watches the boys: from the middle of the line Alex Pirera son of the *Questore* ("Police Chief") makes two steps.

"Anybody else?"

Another boy steps forward, tentatively from the right end of the line. It is Chiesa, the son of the very popular butcher of S. Ambrogio, near Velate, who is making a fortune by selling meat on the black market to the Sfollati. Obviously, his father needs to keep good relations with the authorities and everybody sees that.

Gallini, between furious and desperate, as he must break the news to his superiors, mumbles:

"Two out of thirty!?...They'll kick me out! Perhaps I should have mentioned that new Balillas will fire real guns".

In the porch, where a cane armchair has been positioned at a favorable angle with the daylight, Dentist Tarzilla Scuri opens her bag of suspiciously shaped tools. Mimma holds Franco's hand tightly, while the boy, wide open-mouth, squeezes his eyes shut. In twenty minutes both second incisors that had grown in the middle of the palate, are extracted, with no anesthesia.

For Scuri, the labor has been intense, but the decision to do it had been even harder. She declares:

"Because of the War, I acted against the rules of Orthodontics...I was under Force Majeur ...but I am glad to have done it, particularly because, in this case, the canines, which will take the

Fig. 66 - Dr. Tarzilla Scuri

second place in the arcade next to the incisors, are strong and rounded: I promise, he will not look like a vampire!"[109]

[109] Like many "progresses" in Medicine, Orthodontics is full of rules, that students must learn and apply, even if they go absolutely against common sense. In my case, Dr. Scuri should have pulled two perfectly healthy canines and begun moving the two incisors from the center of the palate toward the outer arcade. It would have meant two years of pushing and pulling in my mouth, with the same final result: the lack of two teeth in the upper dental arcade. It took a war to convince Tarzilla to make the right decision: leave the canines alone, extract the crazily positioned others and call it a day.

Later in the day, the Manor's' children come to inspect the two dark holes in the middle of Franco's palate. They are impressed.

Fig.67 - Franco shows the holes in his palate

The Manor population increases: the town authorities bring in Mr. Maccapani, the Stationmaster of Varese, whose home has been damaged by an explosion (allegedly, an anti-fascist railway sabotage). Maccapani comes with wife and two sons, Walter, 13, and Francesco, 11. Mr. Maccapani is a member of the Fascist Party, and the boys, are fans of the Neo-Fascists Repubblichini, who swarm Varese with hand-grenades at their belts.

The Celadas in the meantime have been taken into the confidence of the Clerici family, and know why Edoardo, son of Emilio and brother of Giulia, is not around: he is an active Political Agent of the out-lawed Christian Left, and dodged the Fascist Secret Militia to flee to Switzerland last year. His activities in the Resistance and the risks involved make him a hero for his younger relatives and their friends. Among them, Franca shows already the seeds of a liberal vision and the passion to pursue it in politics. Franco will learn a lot from her: all his life, he will side with the Radicals instead of the Radical-Chic.

With this background, it is no wonder that with the Maccapani there is "Cold War" from day one. Franca and Franco often end up in catfights against Ettore and Carlo, at times ending on the floor.

Fig. 68 - Kids fighting over politics

The younger boys, Gianni, Alberto and Luca, acting out of pure gut, come up with provocative routines, like calling them a distorted name: "MARKAPAZZ!" (The word "Pazz" sounding like "Crazy") and running like rabbits at the first sign of a reaction. Another time, Gianni and Luca begin parading, with a red scarf around their neck, which is the insignia of the Leftist Partisans (later, Gianni will say: "We just wanted to see what would happen"). What happens if you agitate red tissue under the nose of the Black-shirts? This time, nothing much, except charged epithets, but risks exist.

Franco feels protective:

"Gianni, you fool, take off the scarf! Don't you know the red are Communists?"

Gianni shrugs: "So what?"

In retrospect, he was right: eight-year-olds are not yet responsible for their colors.

At school, the Latin class is disturbed by something happening in a desk of the last row. Under the desk, Ramponi is showing an object to his neighbor Spinelli. It is a gun, and, as to justify this possession, he displays also his *Red Star Card* as *Partisan Dispatch Rider* of Valganna. Other boys, nearby, begin stretching their necks to get a better view. Don Mandrini senses the decreasing attention on Ceasar's *De Bello Gallico*, and, from the expression of Spinelli's face, he understands that there is no time to be lost.

He shouts, loud and calm:

"FIRE DRILL! Everyone stand up and follow me to the courtyard! Let's go!!"

The operation begins orderly. The Don has his eyes fixed on Alex Pirera, the Police Commissioner's son, who wears his usual half-bored expression. He did not notice a thing. Carlo and Franco see that their teacher has white knuckles from the tension of tightening his fists.

The Don prays to himself:

"My God, I thank Thee for this close save…next time please, remember, Thine Priest and Slave can handle show-offs, armed Partisans, Policemen, …. but only one category at a time!"

Frau Hirsch comes to our home on Wednesdays to give us lessons. This time, a cold early January afternoon, she is late. She comes by train; she may have missed a trolley at the train Station. When she shows up, she is agitated and goes directly to Mimma:

"Madam, I am sorry, when we reached Varese, the Black Brigades and the SS were patrolling the station. All travelers had to show their papers…I have no valid papers….

Mimma is flabbergasted:

"What do you mean? You can't…in times like these…"

"I slipped back into the train and hid in the bathroom... I waited until the soldiers finished their rounds…I was lucky".

"Frau Hirsch, this is too risky, for you!

"Yes, I know, but I need to work…"

Mimma, eyes wide open:

"They may catch you! What will you do if they catch you!?"

Frau Hirsch speaks calmly:

"Oh no they won't. If they make one step towards me, I'll jump on the tracks and place my foot on the third rail".

Alice Clayus Longoni relies on her beloved friend Mimma to survive amid the intricacies of the Italian language: while in German most difficulties of expression can be solved by better specifying meanings through simply combining two or more words together, these Italians have different words for each meaning, and sometimes, completely different meanings are addressed by similar words! Foreigners need to be helped in these cases. Franco will always remember Alice's conversation, among the ladies, suddenly stopping in the middle of a sentence, and her frantic cry for help across the room:

"Mimma! GOMITO or GOMITOLO?" ("Elbow or ball of yarn?").

The two women continue to see each other after Mimma moved to the Manor. But Alice's position among the Sfollati, becomes unique, and demanding with the German occupation: a German, educated, non-nationalistic, non-Nazi lady can do and did a lot in support of the Italians and those non-Nazi Germans, who happened to be put in a difficult and hateful role. Obviously, Mimma, and the other ladies, Quadri, Vandoni, Franchini are close friends of Alice, even if they don't take part in Alices' German social gatherings. For instance, periodically, Alice invites groups of German Officers, some of them old Clayus family acquaintances, for musical evenings, where she plays piano and sings Lieders with her delicate soprano voice.

Fig. 69 – Alice Clayus Longoni plays the piano

Some of the invited are Officers of the Prussian military tradition, ashamed of Hitler and, deep inside, furious. One of them, Commander of the *Varese-Land*, is the nobleman Colonel Von Braus. Many are young officers, away from their families for a long time, perhaps forever.

One day Von Braus calls Alice from his office:

"*Gnaedige* (gracious) Frau Clayus! Times are hard, and I will appreciate your opinion on a delicate matter. We received information, from an unorthodox source, that Dr. Taborelli, whom you know well, is hiding non-Arians in his private ambulatory. Before Security launches an investigation, I want to know whether you think this accusation is plausible".

Alice gets the coded message and answers:

"Dear Colonel, Taborelli is a splendid person and an excellent, dedicated doctor! I am absolutely sure he will never act unlawfully or dishonestly!"

Then she picks up the phone again and dials Mrs. Taborelli. The next morning Von Braus okays the inspection of Dr. Taborelli's rooms in the basement. The SS cars start with a roar. No Jews are found.

Uncle Gino. At dusk, a military car is admitted through the gate of the Manor and proceeds towards the porch. An Italian Captain in field uniform and a heavy coat comes to see the Celadas; being Saturday night, also Guido is at home. They had lost track of uncle Gino, Guido's younger brother since he was assigned to the ARMIR (Italian Expedition Corps to Russia) as *Ufficiale di Collegamento* (Liaison Officer) three years ago. In the room with the bookcases, the family sits around him and uncle Gino begins:

Fig.70 - Uncle Gino Celada

"What saved my life are my languages, German, English and French: they put me in Communication, away from the front, in Polish or Ukrainian Railway stations, to manage the trains of supplies, pass orders from the Germans…. a terrible life but in heated buildings with bad food, but food. Very few will come back from the front. They were one million. I am here on permit, to visit the family for the first time. I hope it will be too late to go back, my impression is the war will be over in a matter of weeks…This conclusion was obvious already three years ago when we were sent to conquer Russia no less!" He shakes his head:

I am glad to find you in good health, tomorrow I will be in Borgio, and I have word that my own family and the grandparents are fine… I have something else I have to tell you, but I see Gianni's eyes are shutting down…. Please let the children go to bed, we'll chat some more later!"

Gianni kisses good night and is gone, but Franco would never quit.

The daylight is almost gone, Mimma forgets to turn the light on. Uncle Gino resumes:

"I must share with you something incredible. In Germany there is more than the war going on: Hitler and his gang are trying to eliminate the Jews of Europe. Many suspected this…I was in a position to witness the annihilation of people on the basis of their race …In my stations I have seen the endless "cattle trains" with locked doors on the way to the Concentration Camps in Poland and Ukraine. I have seen cargo trains carrying suspicious toxic substances to the same camps. I have spoken to the drivers of special "modified" ambulances with a special routine: as soon as eight "patients" are seated in the rear cabin and the doors are locked, two men enter the driver's cabin, turn on a switch on the ceiling and drive the eighty or so kilometers to official destination. The drivers get out, open the van from the rear, and discharge the eight corpses in front of the Crematorium: the switch has delivered the right dose of *Zyklon B* gas, a product of IG-Farben."[110]

Uncle Gino looks exhausted. He murmurs:

"You are the first I speak to of this, I could not bear it alone…"

He embraces Mimma, Guido and Franco, taking some time to look in the boy's eyes. Then he walks to the vehicle waiting for him, and he is gone.

---

110. *Zyklon B* is the toxic gas used in the Nazi Concentration Camps, produced by IG-Farbenindustrie (See chapter 11, I.G. Farben).

April 25, 1945: LIBERATION DAY! The busiest and most exiting 24-hours ever experienced by the Sfollati.

7:00 a.m. - Milano. The *Comitato di Liberazione Nazionale* (National Liberation Committee), or CLN, launches a coordinated convergence of Partisan Columns, from the outskirts, to the center of Milan. General strike and insurrection outbreaks freeze the city at all levels. It is a great success. The workers, instead of staying home, occupy the industrial plants and protect them from the danger of "Burnt Soil", temptation of the German army on the run. Allies are still in Bologna.

7:00 a.m. - Milano, Piazza Lavater. At home, Guido throws his Fascist lapel pin[111] into the toilet bowl and sighs with relief. His next move is to pick up the bike he kept in the cellar for this occasion and ride all the way to Velate. He cannot find the key to the cellar nor a locksmith in today's confusion. Plan B: He will walk. He has learned something in WWI: he walks to the fridge, finds the butter, smears it thoroughly on his feet, puts on heavy socks and shoes, a white

111. A hail of clattering pins fell into Italian toilet bowls, most of them followed by a flush, lest any of them had tried an impossible rebound. They were thrown by most honest-to-God non-heroes yet not Fascists, who would not dare to leave their jobs and stop feeding their families, like Guido, by those who had actively participated in nationalistic reactions against red activists at the end of WWI, and even in the March on Rome, like Alberto, and left in spite at that very moment, disgusted by what they saw and understood. The two old friends were in different places on the Day of Liberation, but their clatter and flush occurred probably at the same moment. Intellectuals had had the impulse of breaking up publicly in 1938 when the Regime ruined the University and shamed the nation in by approving the Racist Laws. Some did, and left the country, but many curbed their impulse, hesitated and finally decided to wait the Liberation for the big throw, followed by the flush...

handkerchief under his hat, and is ready to go. He will reach Velate the afternoon of the 26th after several adventures, getting occasional short lifts, hiding sometimes behind road banks, when some stray planes shot machinegun rounds. He picks up a bullet from the turf, for Franco's collection.

Fig.71 - Guido walks from Milano to Velate

9 a.m. - Varese. The eighth graders all arrive to school, having left their home before the news of Liberation spread like wildfire, and are surprised to be met by Don Mandrini in front of the gate on a hired bus with open doors. The Don calls them in:

"Hello Class, today I invite you to an excursion to the Olona River Springs, picnic included! We have a bag of *panini* here, courtesy of the school. Not drinks, but the Olona River water is the purest".

When everybody is on board, he explains:

"Liberation is ongoing, which is very good news, but during political events of this type anything crazy can take place and it is wise to stay away as far as possible from the epicenter of an earthquake. By the way, your families have been notified through the Catholic Parish network".

The students feel well taken care of. However, had they been given a choice, the majority would have chosen to watch the revolution. In a sudden epiphany, they realize that neither Pirera nor Ronconi have come to school today.

The little valley in the woods is easy to reach and completely at the class' disposal, which starts a competition of building dikes, channels and artificial lakes, like beavers. Soon all shoes are wet. As expected, the prize for "Best Dyke" is won by Carlo, offspring and probable heir of the Vandoni and the Gandini Dike Builders, Inc. The Olona River spring is shallow, narrow and without bridges. Thus, another competition begins: to reach the opposite bank by jumping from boulder to boulder and back.

Franco obtains the best time in the first leg and is sure to win: in the second leg, trying to do even better, he slips and falls in the chilly water two-foot-deep with a splash. The class almost dies laughing. He doesn't mind. It is his "Baptism of Liberty".

Fig. 72 - Franco is "baptized" on Liberation Day

9:30 a.m. - Milano. The Archbishop's Palace. Cardinal Ildefonso Schuster, worried about the chaos of a civil war among Italians, most of them Christians, summons Mussolini and General Cadorna (Commander of the Resistance) urging them to stop fighting and agree on a deposition of arms.

10:30 am - Mussolini approaches Schuster, marching in with his entire high Command (*Stato Maggiore*). However, General Cadorna is traveling undercover and, therefore, cannot be found. Mussolini has the expression of a broken man. Schuster offers him a glass of *Rosolio* (a sweet Italian liqueur) and tells him that if he surrenders, a "room" in the Archbishop's Palace will be an extra-territorial harbor for him until his position can be assessed by a proper legal authority. This may not save a man that singlehandedly has dragged his country into a bloody, senseless war, but it certainly could be considered a chance for a legal trial.

11:30 a.m. - Countryside. On the road, Guido is making good mileage, considering the rather wet weather. He gets a short, but welcome, lift on a chariot, near Saronno. The walkers are many, which is a complication. A couple of planes scare their pants off and they hide in the ditch along the road hearing bullets bouncing on the asphalt.

12:00 noon - Velate, Piazza of the Church. A Red, white and green flag is planted on top of the stairs of the church.

1:30 p.m. - Milano, the Archbishop's Palace. General Cadorna arrives with a few members of the CLN; the two sides are about to begin the session, when Marshall Graziani, Fascist Military Commander in Chief of the *Repubblica Sociale Italiana* (RSI), stands up and addresses the Cardinal, shouting:

"Is it true that Your Eminence is treating with the German *Kommandatur* in a convent of this diocese"?

Cardinal Schuster fails to provide a denial. Mussolini is flabbergasted. He stands up and erupts:

"The Germans! This is how they treat us! Like shit! And you, Cardinal, open talks with them before calling on me! Enough of this! How low do you think you can drag me?"

He turns away towards the monumental stairs mumbling:

"And Schuster tells me I should think about my sins!!".

Still furious, he is gone. Many will be looking for him, with opposite aims, everywhere in Milano. Nobody will find him. He will be shot near Como, three days later.

4:30 p.m. - Velate. Two partisan cars go by, one beaten-up Cabriolet with two men sitting on the front fenders, holding automatic guns, the rest of the crew singing the famous partisan song *Bella Ciao*, and addressing a timid population through a megaphone:

"You are free, after twenty-three years of dictatorship! And we did it!".

A second car, a wagon with dark windows, is directly behind it. Franco and Carlo catch a glimpse of this scene while getting off the bus on the way back from their Olona trip.

Fig. 73 - Partisans arrive in Velate on April 25, 1945

5 p.m. - Velate. After a day of news, half news and uncertainties, two cars with red and tricolor flags reach the square and park in front of the church. From the first one, a large vintage open convertible, two partisans, hop off and land limberly on the turf: first, they point the *machinepistoles* to the crowd that is gathering around, then suddenly to the sky, fire two rounds and shout at the top of their voice:

"Hail Freedom!!"

The crowd responds with random applauses.

Carlo and Franco join Franca and the other Franchinis, who are looking out through the Manor's gate, and have a clear view of what happens.

From the second car, a dark, battered sedan, descend two other Partisans, elderly, with red insignia and civil attires, crossing the kids' field of view from right to left. A few steps forward, now out of sight, they stop in front of the Manor's reception; disdaining the bell, they bang their gun butts against the heavy wooden door and command:

"Open Up! We seek Maccapani!"

Natalina lets them in, but forgets to say: "*Benvenuti Compagni!*" ("Welcome Comrades!"), the salute she has been rehearsing for this occasion and escorts the visitors across the first court, where the Sfollati children have clear sight again. For them, and perhaps also as for Natalina, the antipathy on political grounds is replaced by something deeper, the sudden fear of what may happen to a man and a family, who have been living as Sfollati a few meters away from them.

The two partisans reappear with Stationmaster Maccapani between them, reach their car and start a siren. In a few seconds everything in Velate looks like nothing occurred, except for a wife, with a pale face, who stands at the door of an apartment and will not move.

The three days that follow are punctuated by the returns of most fathers, who were away at the magic moment of Liberation, soon completed by the appearance of the Allies with their tanks and jeeps, and girls throwing flowers.

Carlo's father, Pino Vandoni, arrives via company van from Val D'Ossola in the Western Alps, where he was building a dike despite daily fights between German troops and Blue Partisans, who ruled the valley; Guido, with his half-scarecrow-half-explorer attire, arrives by foot from Milano for the joy of Mimma and the children: and, finally, Mr. Maccapani at the end of the third day, is released from prison, unhurt, but with an icy face.

The entire Manor population is relieved. Somebody even thinks the worry was exaggerated, since the Stationmaster had never been accused of anything, except of being a Fascist…

In the following weeks, with the news of the capture and brutal execution of Mussolini, his lover, his faithful and even his unfaithful…the Sfollati see the political hatred raise its ugly head to unbelievable heights. This is what occurs in civil wars, when revenge, interest and petty neighbors' feuds mar justice's retribution[112]. Franco and friends begin to lose their innocence, as their unspoken fears turned out to be more justified than they had suspected.

---

112. It is a accepted truth that civil wars are the worst, but what most Sfollati have learned, and, I bet will continue to stick to, is that there are oppressors and oppressed. Their engagement of the Italian Resistance may not be perfect but cannot be assimilated or even compared with those who curbed liberty for up to twenty-three years. All the young Sfollati are now old, but all remember witnessing the recovery of Freedom as the most exciting prize concluding their most educational tree years.

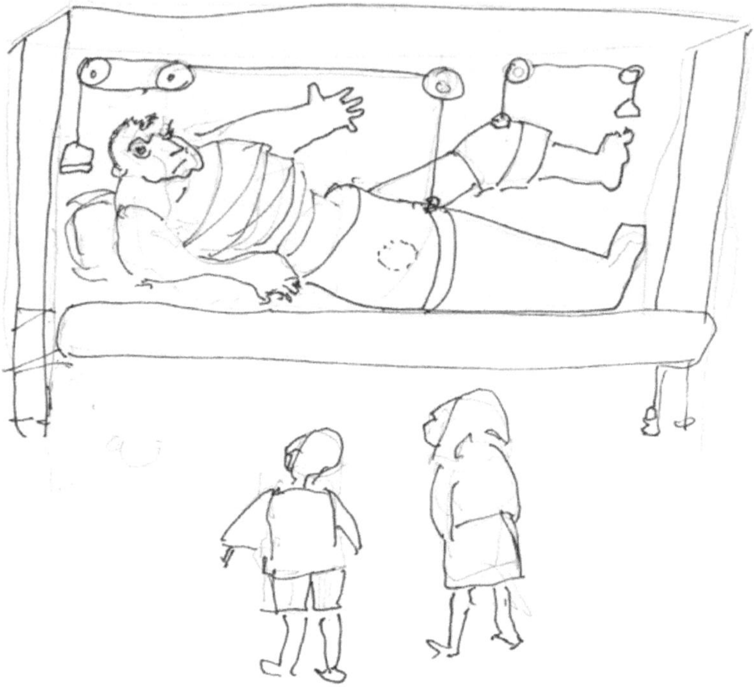

Fig. 74 - Ingegnere Alberto Longoni after the train crash

In the Fall of 1944, two FS trains going in opposite directions collide frontally on the Milano-Varese line. The trains penetrate into each other causing a deadly disaster: three casualties and 62 wounded. The wounded are taken to the Hospital in Varese with an exemplary operation of recovery. The majority of the passengers were on the northbound train and most of them were *Sfollati* around the Sacro Monte area. One of the victims is a 19-year-old student of Medicine, who spent his first day on the premises of the Faculty in Milan and who saves his father's life with his own body. Many of the Sfollati are present in the scene since most of them commute regularly to Milan. Alberto Longoni suffers broken ribs, a broken femur and a piece of window glass planted in the opposite knee. Guido Celada and Adele Vandoni are sitting in the last seats of a car facing each other, when Adele, suddenly airborne, is projected towards Guido, who catches her and cushions her fall with his body. Neither of them suffers more than light bruises and will give a first-hand report of the accident. There are no hints about the cause, because the two engineers appear to have left the scene. Varese's Stationmaster, Mr. Maccapani, and the Police Commissioner Pirera are on the site of the accident and Pirera declares to the Press that: "If an act of sabotage is uncovered, the Republic's repression will be pitiless".

Fig.75 - Natalina places flowers in front of Stalin's portrait

15

Hota Hota

New York is, from its origins, a cradle for diversity, which sets it apart, when the English Colonies reach the fatal decision to gain independence from the old Motherland, infusing the rebellion with patriotic fervor. New York is first *New Amsterdam*, even if the Dutch are easily outnumbered and the European mark of "trade instead of conquest" will never be erased. No wonder, in the Independence and in the Civil War, New York stands out, statistically for its resistance to the Union's Draft[113] to the dismay of its more fervent cohorts (New Englanders, Philadelphians, Virginians).

When General Washington, with his scanty new army, has to retreat from Manhattan in the first serious battle against the British in 1776, he is just moderately sorry for New York being burnt to the ground for tactical reasons. He writes to Hancock, speaker of the Congress:

113. At the outbreak of the Civil War, New York grudgingly supports the Union cause and serious riots break out in opposition to the Draft.

"I did not sign that order, but the Divine Providence has done for us more than I could have asked for (25% of the buildings, including Trinity Church, went to ashes)." 1855 sees a surge in arriving ships at New York harbors, carrying emigrants from troubled Europe. The city has space and the potential for growth, but the recovery from the latest fire has been slow. In this crisis, it acts according to its soul: instead of building walls, like a future President will promise to do[114], or reinforcing the wall at Wall Street, it takes *Castle Lincoln*, (the landmark fortification at the tip of Manhattan aiming its cannons at any vessel approaching from SE to SW), declares its military function obsolete, re-baptizes the fort *Castle Garden* and repurposes it as a welcome center and emergency shelter for immigrants and refugees landing at the nearby harbor. The facility will function for 35 years, until both sanitary and bureaucratic concerns obliges foreign passengers to be screened on Ellis Island, before setting foot on the mainland. The total estimated number of passengers stopping at Castle Garden is over eight million, although that sounds high.

The subway train n.4 takes me from Union Square to Bowling Green in four stops. When I exit the station, the Castle is in unobstructed view on the far side of the lawn. I walk around it and inspect what remains of the 1802 structure: the stone wall all around, with 16 cannon holes aiming at the bay. I stick my head into some of the holes, which open at ground level because of a landfill remodeling of the shore. There are no cannons any more. There might have been, when my family arrived in 1855, among the first of the eight million, certainly not expecting any kind of special welcome prepared for them, a few steps away from the shore.

114. See: Trump, Donald , *Collected Campaign Speeches*, 2016

Fig. 76 - The Bay Harbor of New York with Castle Garden in the forefront, 1855.

They are one adult, Giovan-Battista Bonino from Chiavari, on the Ligurian coast, and his two sons, Francesco, 12, and Giuseppe Giulio (the JJ or Hota Hota, mentioned in Chapter 1, *Porcini*), 7 years old. It is not one family, it is one half of a family, sliced in the middle by gender: Paola and the two daughters remained behind. Even the immigration officer behind the desk is alarmed:

"Mr. Bonino, what will you do? We can keep you here for a few days, but the children are so young..."

"Comandante…Sir! I must find connections, I have addresses. They will find a job for me, I will scout the city during the day…don't worry about the boys: they are good boys and able to work!"

I don't know what the Comandante said, and I don't know whether this ancestor realized how lucky he had been arriving just after, instead of before, Castle Garden was opened for immigrants. All I know is what I learned from several interviews with uncle Rinaldo, JJ's eldest son, who heard the story directly from the horse's mouth: the boys succeed in staying at the Castle, and they also help the custom officers with their inspections of the cargoes on the peers. The officer, using a wooden cone and a hammer bores a hole in one of the burlap sacks amassed on the pier, collecting a sample of the content. The boys have to repair-stich the hole. If contents look edible, they may take a second sample for the pot. Giovan Battista keeps coming back every evening shaking his head…He is no winner. Who knows how many cold shoulders and closed doors he has found, and who knows if there was spirit of comradery between Italian immigrants coming from different parts of a country which is still not united. One evening he tells his sons:

"New York is difficult, no job for me…but I have a real, faithful friend in San Francisco. We must go there. There is a caravan leaving next week, we have to prepare ourselves"!

"But, dad, that is on the Pacific Coast!"

"Yes"

"How do we get there?"

"By foot. How else?!

Francesco and JJ finally accept this new challenge; they are not new to walking: departing from Chiavari, it had taken the trio twenty days to reach the harbor of Marseille, a distance of 430 km., where they had boarded a New York bound Sailing Packet Ship[115] (or was it a Clipper[116]?).

Fig.77 - Vieux Port (Old Harbor), Marseilles, France, 1890.

Leaving New York, they probably do not realize how much more land lies ahead: 4000 kilometers, ten times the Marseille distance, more than 200 days walking. We, uncle Rinaldo and thus I, know very little about the trip, except this dreamy memory of a night in the wilderness: dark, with fires around the camp, the sounds of wolves in the distance, the children tucked in blankets trying to sleep and Giovan Battista with other men, adding wood to rekindle the

115. Commercial ships, carrying mail and wares, with Spartan accommodation for passengers.
116. *Clippers* are the fastest sailing ships, with 3 or 4 masts, looking like overblown Schooners, and typically employed on the routes of the tea.

fires. That is all. Except the fact that they made it to San Francisco, defying all odds, and met Giovan Battista's friend.

I say:

"Uncle Rinaldo. This sounds crazy! What proof do we have of this incredible story?"

He smiles, sweetly and says softly:

"Francesco is the proof."

Uncle Rinaldo is 81, ailing in a hospital bed in Genoa. What he means is that in S. Francisco Giovan Battista did not find a job. He left the 12-year-old son in custody of his friends, since they could use the help in their general store and, with JJ, embarked on a sailing ship back to Genoa, after crossing the Equator, Cape Horn, Madeira and Gibraltar.

Giovan Battista thinks: "Did I make the right decision? Will I see Francesco again? Will he be lost to the family? What will Paola say?"

Time goes by. Time, for the father to try to forget the nightmare of his endeavor. Instead, the young JJ does not want to forget. The unsuccessful attempt is a formidable challenge for a boy with his genome.

At age14, he begins to explore Genoa's piers, the traffic, and the seamen's market. In 1862, when he is 15, he is hired as *Cambusiere*[117] on a sailing ship on the Cape Horn route, a *Packet Ship* carrying payload and passengers in open spaces and a few cabins. JJ distributes the daily rations of oatmeal, dry fish, salted meat, water and biscuits and it is up to the passengers to cook, or, rather heat the food on an oil flame.

117. Galley clerk. The crew member in charge of the daily distribution of food rations to passengers.

Fig. 78 - A Brigg on the route Genoa-Callao in 1862

The crews for sailing ships are large, they sleep in shifts. The hammocks hang in the middle deck, and through loopholes, the cargo in the hold can be seen. The worst storms are those around Cape Horn, this time true to its severest reputation.

Fig. 79 – A brigg caught in the tempest

At nighttime, JJ has trouble sleeping: the ship is keeling, and pitching through the waves; when it comes about and begins the next tack the keeling changes sides fast, the hammock swindles gently, while the cargo, in bundles and cases, breaks free from its cables, and slides on the floor, from Port to Starboard and from Starboard to Port in the next tack. The menacing sound and the shadows seen below are fixed in JJ's memory.

When the ship drops anchor at Callao, Lima's harbor, its sails broken and with water in the hold, the captain must throw out most of the spoiled cargo.

A part of it are shoes, once, a valuable item. When JJ sets foot on the quay, he sees a little mountain of dark damp, smelly shoes in front of him. They are garbage, ready to be thrown. It takes thirty seconds for him to make a decision:

"Capitano! The rotten shoes! Give them to me! I'll take care of the disposal! The captain knows the boy well, after two months of having him on board. He smiles and shouts back:

Fig.80 - Callao Harbor, Lima, Peru, 1866.

"They are yours, JJ! Good luck!"

One week of scrubbing and shining shoes on the pier, is interrupted only by fights with gangs of local boys who come around laughing at him and calling him *Gringo!*[118]. One of the family stories has it that JJ was floating in the harbor, thrown in the water by a gang the local boys. When he asks to throw him a rope to climb back to the pier, they oblige, but, when he is half way up, they let go again, shouting: "Gringo! Gringo!"

The incident is not going to change his mind. Astoundingly, he succeeds in selling the reconditioned, Made in Italy[119], shoes, which are of good quality even if wet, for some much needed cash. The

118. Derogatory term for "stranger" in Spanish.
119. Italy has been united for only one year, since 1861.

news spreads around town. JJ stays in Lima, his first job is "Boy in a shoe shop". Then assistant, then owner of the shop, then, of a chain of diversified shops. Very soon JJ realizes that he needs collaborators, and how difficult it is to find somebody he can trust. He sits down, writes a letter to an address in San Francisco, takes it to the harbor, puts it in the box for Northbound ships.

The enthusiastic answer arrives by mail, and his big brother in person arrives in Lima in less than two months. Francesco proves to be a good accountant, and he admires the higher brilliancy of the younger brother. Now JJ can expand his commerce.

The city of Lima is a crucible of races and ideas, and JJ has a knack for understanding their taste and their needs. He designs stylish footwear for the ladies and inexpensive jewelry. The drawings are sent to Paris, at that time more than ever the *Ville Lumiere*, where the items are manufactured and shipped back to Lima to be sold in the *Bonino Novelty* shops. The cycle takes four months: an impossible waiting time today, but "a very short delay if you beat the international competition", says JJ. He becomes rich. It will soon be time to marry: it is certainly not too soon!

The indomitable adventurer reveals that he is still mama's boy when at 30, he asks Francesco to run the business alone, while he goes to marry a girl from Chiavari. Mamma Paola expected nothing less, and she is ready to take charge of the ritual match-making. A look at the photograph of JJ with his sisters is sufficient for Paola (born Brizzolara) to have the situation well under control.

A very good clan to marry into in Chiavari is the Devoto cousins: two families, living in the same building. They have several daughters: two of them of similar age are allowed to walk on Sundays up and down the *Carruggio Dritu*, the long, elegant, central street in Chiavari. If you happen to like what you see and want to have a second look at today's lineup, do not come around the Carruggio next week, because each pair walks every other Sunday. Daughters are precious: they should be shown but not too much. JJ, nicknamed *The American*, is invited to tea very soon. Somebody introduces him to Virginia Devoto, and the next day he asks for an appointment with her father, Emanuele. On this second visit, as he is following Mr. Devoto to his studio through a corridor, he has a passing vision of a very young maid, scrubbing in the kitchen. The two men begin to converse:

"Dear Mr. Devoto, I asked to see you because I met your daughter Virginia.."

"Virginia?!"

"Well yes, I think I like her, and I came from abroad to marry."

"My dear Mr. Bonino! I am sorry, there is a misunderstanding! Virginia is already engaged. She is due to marry the honorable Mr. Besio, from Genoa."

"Really! Does this mean that I should have been introduced to another daughter of yours? Is this the only misunderstanding?"

"Yes and no, Mr. Bonino! There is another daughter, Emilia, but she may be too young for you, a mature man, with a position, with experience… I hesitate proposing her! She is still sixteen. You might have seen her hiding in the kitchen …"

"I have, Mr. Devoto. And let me be totally frank: "*A piccin-na, a me piaxe ancun ciu` tantu!*" ("The little one, I like her even better").

This is a thrilling start for a love story. But at the moment there is no love, only doubts and desperation. Emilia, not yet 18, laments:

"This man has a beard and a mustache; he is even four years older than Shu Besio, my older sister's fiancé. And that is not all! Now everybody wants a quick answer, and I must make a decision!".

She is right, she cries a lot. It takes the deed of another mother to reach an agreement where feelings and conscience are protected.

Cristina Arboco', mamma Devoto takes the matter in her own hands. She has a clear mind:

"We want a good marriage for our daughters. We will have a double wedding." This will not only be a memorable event in Chiavari, but a traveling honeymoon of the two couples is important: the proximity of an older sister will reassure Emilia. And in order for this story to have a happy ending, I will have a secret meeting with JJ at 5 a.m. on the day of the wedding, September 23."

When the date of no return comes, the meeting of groom and mother-in-law takes place. Cristina has many recommendations for JJ and only one request:

"This is the condition for us to give parental consent for a girl of 16…" She opens a bag, extracts a Bible and pronounces:

"…Do you, JJ Bonino, promise that you will never take my daughter Emilia to America?"

JJ makes a surprised, then solemn face, places his right hand on the Holy Bible, and utters:

"I do"

Here is an unexpressed thought of JJ's:

"Arboco'! I thought I had had more than my share with mommy Brizzolara!"

The double honeymoon to Torino, a beautiful but not particularly joyous city, is not amusing. The Besios do not want to chaperon all the time, the Bonino are still unable to relax. In short, Emilia and JJ are sleeping in separate rooms. When the allotted time is over for the Besios, they go back to Genoa happy, but with worrisome news for the families.

JJ asks Emilia: "Do you remember I mentioned my drawings I have to send abroad…?

"Yes, you said to Paris, I wouldn't forget that."

"It *is* Paris, brava!"

"And Torino is just on the railway path to Paris. Thus, by adding two and two, we could take the train, make the contacts for business, and spend some time visiting the most famous city in the world. And we would be only one couple, instead of two. Would you like that?"

"Julio, I cannot decide...you hear so much about Paris. Is it really possible?"

"Yes Emilia, and here are the tickets I booked yesterday. The hotel valet just brought then up!

A little kiss would be the right conclusion of this dialogue, but I am not sure it materialized, thus I say nothing.

The honeymoon in Paris is one of the cunning intuitions of JJ. The city, half way to its bloom, has most of its attractions, even if Mr. Eiffel's Tower is still being built, and *Le Metro'* is still fifteen years in the future. The couple finds parks, boulevards, palaces, the *Arc de Triomphe* and theaters, music, feasts of incredible scale, all reachable by an incomparable network of electric trolleys.

They must have shopped for elegant garments, simply by observing the photographs of the couple and their individual portraits made by the *Atelier Petit* printed in ochre and brown. The photos of Emilia are particularly attractive.

It is a pity that no portrait was commissioned to one of the painters active that year in Montparnasse, the still not famous Cezanne, or Seurat, but a photo is certainly more faithful. We are reassured: Emilia is becoming relaxed and happy with "Julio", as she has begun to call him, and this also promises that there will be, in the future, a discrete number of descendants, which is of importance, especially for the descendants.

Fig.81 - Tour Eiffel still under construction, Paris, France, 1885.

JJ must be happy. The end of the month approaches. He has prepared a little, defensive scam, in anticipation of possible future family reproaches. Two or three days later he enquires with the hotel clerk whether a telegram arrived for him. The response is as expected:

"No, sir, and be sure that if a telegram comes in, it will be delivered to you immediately."

The scam begins here:

"Julio, why are you suddenly impatient? From where do you expect a telegram?"

"Oh, from my people in Lima, don't worry, Milia, it is supposed to arrive tomorrow". Another day goes by:

"Julio, please tell me: what happens in case the telegram does not arrive?"

"Well, in that case, we must go to Lima.."

This scam is called: *Il Ratto della Sposa* ("The Abduction of the Bride"). This time Emilia's reaction is mild. She does not shout, nor cry, she does not mention JJ's sworn promise to mamma Cristina. Being a smart girl, she probably suspected JJ's move, even if she will forever deny it. She asks:

"When?"

"There is a steamboat in Cherbourg in three days. We must take the train from the Gare du Nord tomorrow at noon.

Transatlantic travel has just made a big leap, with the advent of steam engines. These liners are not very large, they keep the sailing option as default, but they have cut the voyage time considerably. France is among the forerunners in this world competition. By coincidence, one of these steamers, the *Insere*, has recently carried a special gift from Rouen to New York in her hold: the Statue of Liberty, still divided in 212 crates. Arrival in New York: June 17, 1885. Independently, newspapers acclaim the launching of four new French Liners between 1885 and 1888.

The newlywed board in Cherbourg in late October 1885 and make it to Lima in less than two months. There is amusement on board: JJ is at the player piano. Couples begin to dance at the crossing of the equator. Emilia is approached by an elegant passenger, Charles De Lesseps[120], who invites her to dance. Emilia smiles politely and says:

120. Son and co-worker of Ferdinand De Lesseps, the constructor of the *Suez Canal* in 1869. Father and son attempted to repeat their success in Panama in 1888, but failed.

"I am sorry, sir, I don't know how to dance!"

I don't have many details of the arrival in Callao. They settle in Lima to a comfortable four years. Emilia has a chambermaid and JJ hires a Chinese cook with a long braid down his back, and gives him money every morning to go shopping and prepare two meals a day. It seems that Emilia, with the exception of the brother in law who is of great support, is seldom (or never) introduced to her husband's local acquaintances. The great news is the birth of the first child in February 1887, a girl named Elvira, the first offspring of seven. The seventh is Erminia, my mother, born in Genoa in 1901, the "Mimma" often appearing in this book (See Chapters: *Porcini, Stefania and Sfollati - Part One*). Emilia, however, disagrees with the climate of Lima, alternating cold summers with Lima's fog, made of water droplets quite larger than the invisible ones of Milano. (I was able to compare the two during my trip to Peru' in 1976). Emilia seems to breathe with difficulty; the specter of the *Mal Sottile* ("waisting disease", TB or Tuberculosis), so pervasive at the time precipitates a decision that was already in the plans for the future. The business is left in the hands of the brother Francesco, and later to a nephew, Francisco. In 1890 JJ and Emilia with the 3-year-old daughter embark for Genoa. Genoa becomes their town, even more so since, the Devoto, have bought an apartment in the popular Via Colombo to stay near the daughters. Emilia is happy, a favorite of all Devoto relatives, considered a heroine …

JJ savors the first years of life without work. He observes, and studies the city, which is brewing progress and initiative. Every two months he receives a report with the accounts of his Lima assets, and business. The second year, it is young Francisco who begins to sign the reports.

JJ knows that he needs to have his assets near if he wants to remain in control. When he opens the last report, it also contains an elegant business card of his nephew, which reads: "Count Francisco Bonino, ESQ." This sight triggers one of JJ's momentous decisions: He cannot rely on a "Count" to administer seriously, especially not on a fake Count, roaming free overseas! He books a passage to Lima. In less than one year he sells his chain of shops, transfers the capital to the Bank in Genoa, thanks and compensates former administrators and makes his sixth and last crossing of Cape Horn and the Equator.

Fig. 82 - Genoa, 1890

In the meantime the city of Genoa continues its progress: short of space, it remodels the Eastern limit marked by the medieval wall, and extends it down the banks of the Bisagno river, acquiring one square mile of prime housing locations, builds a *Monumental Bridge* connecting sections of the ancient walkway on top of the city walls, relocates Porta Pila on a nearby hill as a remembrance, and cuts a new wide boulevard reaching from the top of the city plaza, where the Duke's Palace and the Opera theater are located, down under the monumental bridge and continuing straight over the Bisagno: the latter is simply interred. The new street is named Via XX Settembre, in remembrance the Breach of Porta Pia in 1872, Italy's conquest of Roma as its capital, and the end of the Pope's political and temporal power: a good deed, especially for the Church.

The rebuilding of Genoa takes years, and coincides with the passing of the century. JJ is there: he buys the entire storefront of one of the central blocks in the Via XX Settembre. The strip consists of ten stores of various configurations. Since the building is in construction, JJ can design the shops, and, especially, their widows. In his experience, the windows are the most important factor to sell anything. His shops look modern even now.

One block down from the shops and four blocks to the right, Via Frugoni follows the outline of the former town walls. There are only buildings facing south, with windows and terraces in "floral" style and offering unobstructed view of the coastal hills and glimpses of the sea. JJ buys the fourth floor in 1901, constructing and designing rooms and furniture.

Fig.83 - Genoa, Via XX Settembre, 1903

On the shoulders of the giant

It looks as if not much of the original "globetrotting" spirit, or gene, is inherited by the first generation of the Bonino family: a total of five, two boys and three girls, including the "Peruvian" Elvira (not counting two more girls, sadly lost at the age of two and seventeen). Or did they inherit some "not-yet-recovered" fatigue, endured by their parents in the New World? A moderating effect, was certainly produced by JJ himself, who asked Emilia not to wake the children too soon for school, arguing:

"I remember how hard it was for me, how many hours of missed sleep at their age! I have amassed so much work, as to allow my children to rest in the morning!"

In the second generation, there are seven grandchildren: five boys and two girls, and, at least, one of these cousins fulfills the Lima requirement: visiting the place where JJ lived for 25 years. I was there only five days in 1976, found the location one of his properties, but spent more time in Machu Picchu and Cuzco. One thing is for sure: I share feelings with JJ. How could I not, since he is undoubtedly the first Globetrotter?

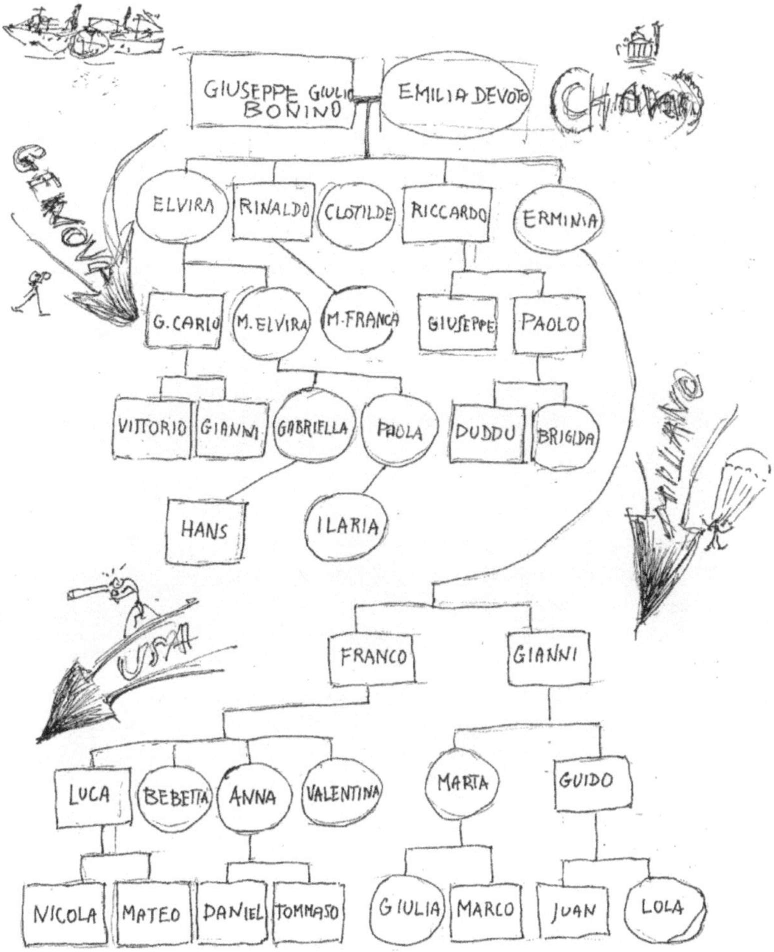

Fig.84 - The Bonino family tree with the different settlements in Chiavari, Genova, Milano and U.S.A.

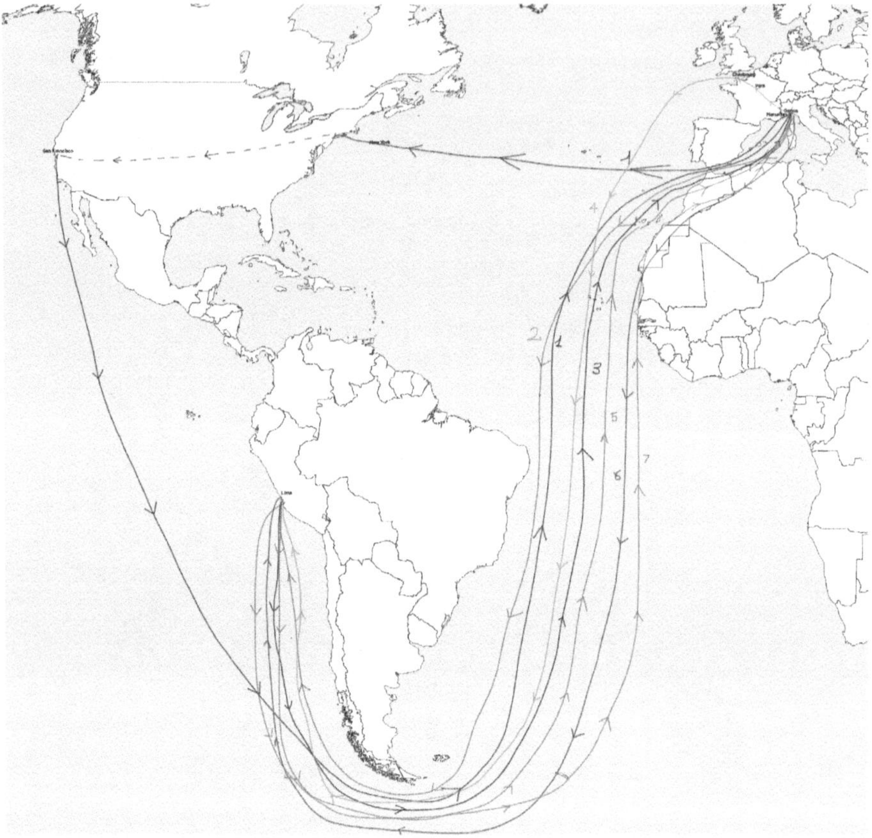

Fig. 85 - The seven trips of Giuseppe Giulio Bonino (Hota Hota) between Europe and the Americas.

List of Hota Hota's trips:

1. 1855 - age 7 - *Genova - Marseilles - New York - San Francisco - Genova*
2. 1862 - age 14 - *Genova - Lima, Peru*
3. 1884 - age 36 - *Lima - Genova*
4. 1885 - age 37 - *Genova - Paris - Charbourg - Lima*
5. 1889 - age 41 - *Lima - Genova*
6. 1895 - age 47 - *Genova - Lima*
7. 1896 - age 48 - *Lima – Genova*

Fig. 86 – Cape Horn and the "Cape Horn Monster"

Cape Horn

Seven crossings of the dreadful *Cape Horn* is a respectable lifetime record even if the last 3 or 4 were on board of steam-powered vessels, decidedly safer than sailing ships because of their ability of choosing their course with more precision, in straits ,amid rocks and groups of islands. The southern tip of the American Continent has the worst winds, storms, thirty-meter-tall waves, sudden snowstorms, icebergs escaped from Antarctica, and the most blinding fogs ever heard of. This was well known as news of disasters were heralded, but did not discourage the attempts, particularly in the early and middle 19th Century, when all carriers were sailing ships. The frequency of attempts, and consequently of victims, peaked under the influence of great and tragic human events: *The Gold Rush* in California (1820); the building of harbors on the West Coast stimulating commercial Trade from New England (1830); the early *European Emigration to South America* (1840); the *Guano Exploitation*, which also caused the transportation of slave workers from China to Chile for a deadly job (1850 and on).

The Cape Horn "Monster" was talked about, but did not influence the travelers, just like Ulysses, in Dante's Inferno, was not deterred from his quest beyond the Herculean Columns. Millions of "future-seekers" will die as true heroes. Only the opening of the Panama Canal in 1914 decreased the number of crossings.

Fig.87 - WWI: Italy is at risk of being defeated by the Germans after the breach of Caporetto. Frantically, the government tries to strengthen security by limiting its citizen's internal movements and Giuseppe Giulio Bonino is forced to use an Internal Passport even to travel from Genoa to Arquata.

Fig.88 - Giuseppe Giulio and Emilia Bonino during their honeymoon in Paris

Fig.89 - Portrait of Giuseppe Giulio Bonino at 60

16

The New World, in Two and Four

In October 1959, the first commercial jets flew over the Atlantic. They were expensive. Antonella and Franco at Malpensa Airport board, instead, on a propeller-equipped contraption looking like the bombers they had seen over their heads 15 years earlier. The monotonous roar of the four-piston engines is difficult to become accustomed to and will last for twelve hours. In fact, there is a strange feature of this piston-roaring beast, that you can't imagine if you haven't seen one. Let me explain with additional details. We plunge into darkness. They give us a so-called meal in a bag. After that, water or tea. Then the cabin lights are dimmed, and most passengers begin to doze off, and perhaps even fall asleep. After two hours I wake up. Where am I? We are positioned in the rear of the cabin on the starboard side. I recognize one window from which I had the view of the right wing. The window is closed by a rolling shade. I feel for the catch. The shade is sucked up. I see…FIRE! TWO FIRES! Two four-feet-long cylindrical flames are gashing out of the exhaust holes at the rear tip of each engine: a scene from hell, illuminating the outside of the cabin. For me an instant of terror. Luckily, I do not shout. An attendant comes by, and says, smiling:

"First flight?"

"By night, and transatlantic, yes, and I did not know the engines go on fire! Seriously, I have a question for you: I was at this same window before dusk and I saw no flames! Why?

"Sir, the exhaust gas has always the same temperature, but you see the red light only in the dark".

New York: Dip and Go

If you are a traveler bound for some far place in the US in 1959 chances are that you will land at *Idlewild*[121], Long Island and spend a couple of days locating the luggage you shipped separately from Europe two months earlier. To do this, you have to penetrate into Manhattan, book a hotel, find in which of the official downtown buildings of the South Street Seaport the trunk is lost, inspect it, clear the customs and pay the shipping company, which you thought you had already paid in Milano. The hotel is in Midtown; unfortunately, your room's window faces the back alley where garbage is collected and garbage vans work, loudly, from 4:00 a.m. to 9 p.m. This is not the best way to visit *Wonder City*, but for sure you can learn a tremendous lot, thanks to the novelty that permeates the New World. The subway we decide to take, to decrease our distance from the trunk, is crude. Iron, all black and unfinished, and has a different feeling than the Parisian *Metro*. On top of all these novelties, our own

121. The New York International Airport before being renamed "JFK" after President Kennedy.

must be counted; Anto's pregnancy, the first of the present generation for both families, extraordinarily healthy, when judged from the result, but full of petty discomforts, marring the normal pleasure of the meals for the mother-to-be. The couple and the trunk continue their separate itineraries, this time by train, aiming at Oak Ridge, Tennessee, in the Appalachian Mountains. The couple descends from a train that re-starts in the opposite direction. It is not really a station, but the end of the rails, equipped with some open wooden platforms. They spot the trunk on one of them. Having assembled their hand luggage, coats and things around the trunk, they follow the instructions received from The Oak Ridge National Laboratory: "Wait at the station: arrangements have been made for your accommodation, and you will be picked up promptly". Not so soon, a tiny black dot, which becomes a car and then, more precisely, a Plymouth 48, appears in the distance.

At the wheel of the Plymouth, geneticist Benedetto Nicoletti from Rome, who has been working at Oak Ridge for one year, begins to discern the group that may correspond to the description he got from the lab. When he stops, he is not sure anymore: they have a gun among their belongings![122] The woman is spitting in a beaker:

"I can't really make them out! Anyway, it is the only couple around".

---

122. Note: the gun is a Winchester caliber 22, 6-shot carabine, designed for marksmanship as a sport. It was bought in Milan as a crazy present from Franco to Antonella before the Oak Ridge plan emerged. It has a wonderful Italian leather case. In the enthusiasm of the adventure it was carried back to the US, fired once, aiming at Coca Cola cans in a deep forest in Tennessee. It was smuggled back to Italy, where it was never fired and lies, still… in a chest drawer. Nobody is proud of this object. In retrospect, it was crazy to give it as a token of love, considering the symbolic, tragic meaning of firearms and, in the US, the polarization around the Second Amendment of a distorted, unacceptable sense of freedom.

Fig.90 - Franco and Antonella, Oak Ridge Train Station, 1959

"*Buongiorno*, and welcome". After the introductions, he explains:

"Virginia White, the factotum of the Biology division, sent me because of the language, and because I am a bachelor and have no excuses! She also asked me to give you these tickets for tonight: two suites by Mozart."

Now is the turn for the newcomers to become astonished:

"From where do orchestras come to this desert?" asks Anto. Benedetto replies:

"You are right about the desert! But you ought not to underestimate biologists... We have an orchestra and chamber music, all staffed by researchers, opera with a soprano, Virginia White, and prose: next spring we will have "LOVE for LOVE", a jewel of the English Restoration…; and the visual arts as well, mostly produced by the wives of the biologists! One out of four wives is a painter."

Anto, an artist herself and a graduate from the Academy of Fine Arts in Brera, Milan, looks at her husband, in horror.

To Live in the Heart of the Uranium Blower

Oak Ridge, as the name says, is in a lowland strip made up of narrow picturesque valleys paralleled by a series of small wooded ridges, an isolated region between the Great Smokey and the Cumberland Mountains. In 1942, the Manhattan Project chose this location to develop and construct the atomic bomb from scratch and in the highest secrecy. The fundamental need was ^{235}U (uranium 235), an isotope[123] of ^{238}U[124], and nobody has ever done this purification on an industrial scale. Two methods can be used for this purpose:

1. Throwing uranium ions into a cathode tube against a screen while subjecting them to an electric field, which will deflect more the lighter isotopes than the heavier. The apparatus resembles an enormous television cathode ray tube. And the

---

123. *Isotopes* are atoms of a chemical element with the same atomic number and position in the periodic table but different atomic weight.

124. ^{238}U is the most abundant form of uranium in nature. ^{235}U is a lighter isotope, radioactive, representing less than 1% in most natural sources. Only ^{235}U is used as fuel for nuclear reactors and bombs.

"pure" product will be scraped after the operation from the inside surface of the tube.

2. The other method requires a heroic chemical reaction: obtaining the hexafluoride salt of one of the heaviest elements, uranium hexafluoride (UFl6), and forcing this gas into a series of so called "Fog Chambers". Here, the rare ^{235}U will be a little faster, and after sufficient runs, will come out "pure". The ^{235}U atom has 3 neutrons less than ^{235}U, but it is radioactive and *fissile*[125]: it is, therefore, the fuel for atomic bombs, based on the process of nuclear *fission*. If hit by a slow neutron the atom breaks, triggering a nuclear reaction.

In the laboratory tests, both methods were hopeful. Industrial savvy would suggest, in this case, to construct a small-scale plant for each method, compare the results and only then build a full-scale plant. Instead, the Manhattan Project, whose priority was speed and the necessity to win the war, decided to launch in parallel both projects in full scale: the cost will be double, but the time will be cut by half. The Fog Chamber won the competition and production began. The construction of cathode tube was discontinued, leaving a mysterious, half-finished empty cube crowned by stacks, abandoned for the next twelve years.

125. *Fissile* is the property of an atomic nucleus to split into two nuclei when hit by a slow neutron.

Fig.91 - One of the Manhattan Project Buildings (1947), which will become the ORNL Biology Division in 1955.

The race with Germany was won with no need for atomic bombs. Einstein's letter to President Roosevelt had been rather on the "better safe than sorry" side: Hitler was hopelessly behind, this time. Did the Japanese need to be burned away for them to surrender? Probably not: as the Manhattan Project scientists had suggested, dropping the bomb two miles away from Tokyo would have been sufficient. However, in August 1945, President Truman decided otherwise and two bombs, one built at Oak Ridge and the other in New Mexico, were dropped on Hiroshima and Nagasaki. Both exploded, and the damages were higher than expected. The Manhattan Project was thus concluded.

In 1946, a scientist of the National Institute of Health with a strong German accent, Alexander Hollaender, was offered the direction of the Clinton Laboratory, comprising a few structures of radiobiology and health care, which were used during the building of the plants in Oak Ridge. The appointment would be subject to the approval of his full program, and a budget would follow. Two weeks later, Hollaender appeared to the Commission. He explained his motivation:

"There is an incredible gap between what we do and know in nuclear physics, and the little we know about what happens when radiation hits the living matter. This is odd, wrong and dangerous. My program is to build a laboratory that studies radiation on all kinds on living beings. It will take a massive cooperation between physicists, geneticists, immunologists, pathologists, and microbiologists. We will study effects and remedies on all living beings, from viruses to mammals, from cells to plants, mollusks, fish… It will not be a normal laboratory. It will be the *Biology Division of the Oak Ridge National Laboratory*. It will begin with a staff of 70, but will easily reach a comfortable 200, which is, in my mind, the *critical mass* for brains!"

The commission must have liked the program, and particularly, the concept of critical mass: a very appropriate analogy for this capital of uranium. There were no limits in the budget.

The Biology Division, certainly the most advanced interdisciplinary, internationally staffed research institute, takes possession of the former building of the mass spectrograph[126]. Obviously, mass spectrographs do not have windows, and in the "cube" neon lights and air-conditioning are always on. There are still large stacks on the roof, designed to suck out and replace the internal atmosphere if something should happen to the mass spectrograph.

When I arrive, six full floors had been added to the giant cube and there is space on top, should a new floor be added if the number of research groups increases. The Immunology Division is a relatively new addition, and, therefore, the Takashi Makinodan group, where I have been enrolled, is on the 14th floor.

During my two "windowless" years in Oak Ridge, I invent ways to measure how sensitive our immune defenses are to radiation and written seven papers, thus becoming a "published immunologist". I also come to know and collaborate with many of the over 200 researchers of the Biology Division, but from the relationship with the five of the 14th floor, personal memories and insights remain with me forever. Regrets? Only one: not being able to play table tennis with my friend and colleague, Peterson, as you will see later on.

Our team consists of: Makinodan, a Japanese-American citizen born in Hawaii, whose family settled in California; Russo, a first-generation Sicilian who communicates better in English than in Italian; Gengozjan, an Armenian, whose family settled in the US after the Armenian genocide occurred in 1915-17; Perkins, a religious scientist, bishop of the Church of Jesus Christ of the Latter-Day Saints (a.k.a. Mormon Church or Nontrinitarian), who takes off every weekend on a mission to visit the new communities of Tennessee; and lastly, Peterson, an excellent African-American technician, greatly valued by Makinodan, but considered an oddity in segregationist Tennessee, where technicians are still strictly white.[127]

---

127. The population of Oak Ridge showed tremendous fluctuations, peaking at 80.000 during the building phase of the *Manhattan Project* and settling down to about 50.000 in the late 1950's. Also the ethnic distribution was odd and not virtuous: the African Americans were relatively few, drafted for basic jobs: in the lab, as attendants in the animal colony floor and, in town, mostly as home cleaners. The segregation was almost total. From the beginning, their own dwellings were confined on a hill, one mile away from the town center. When the town managers thought that the bus service (which was used practically only by African Americans) was too expensive, they simply discontinued it.

Makinodan is stone-faced, calm but reserved. Occasionally, in discussion with his peers, a slightly out of place vehemence in the arguments emerges. Only at the end of my period of collaboration, does he confide his secret to me, which is still not condoned and would never allow a complete blending-in with his co-citizens and colleagues. He was still a boy, living in California in 1942, when Japan struck Pearl Harbor. His family and all the other Japanese-American families were sent to internment camps for three years.[128] They were all US Citizens. They never expected such treatment. A verse from Dante's Inferno pops up in my mind. It is the perfect description for his feeling: "*E 'l modo ancor m'offende*" ("And the manner still makes me sore!")[129]

One day Russo comes to me:

"Franco, I have hesitated to come and tell you this, but I must: I am a Catholic. All my life I had to confront friends and neighbors laughing at me for things like the rules of fasting! Now, you come here, every Friday, and eat your steak or ham sandwich. I will not even ask you whether you are religious, but I assure you that everybody in my America makes this immediate assumption: all Italians are Catholic, Franco eats meat on Fridays, therefore Russo is stupid!"

My reaction of total astonishment gradually becomes compassion, and I explain that, during the war, in Italy the Pope cancelled any rule that would add to the struggle of the population,

---

128. In 1942, by presidential order, 120,000 Japanese-American citizens were forcibly removed from their homes at the Pacific Coastline, and relocated in 10 Internment Camps until the end of the War. Two of the Camps were near the Californian Desert, the other 8 in states further Inland.

129. Dante's *Inferno*, Canto V – verses 100-102. Francesca describes the assassination of herself and her lover Paolo, by her husband Gianciotto: *"Amor ch'al cor gentil ratto s'apprende, prese costui de la bella persona che mi fu tolta; e 'l modo ancor m'offende".*

and then forgot to cancel the cancellation! Russo looks at me, puzzled, so I continue:

"But, I have no problem refraining from eating meat once a week for your sake! I will relay the message to Antonella; besides, her frittata sandwich is the best!"

Gengozjan takes it upon himself to give orientation tours to the newcomers, explaining the mysteries of the Cube, its routines and its rules. He takes me to the animal facility on the top floors of a nearby building, the irradiation units, the Bio-Mathematics & Statistics Division, where I will be able to discuss the optimization of the experimental designs, and verify our graphs, with the editorial office, a godsend place. Here we all learn to write our reports in the simple style of Science, and I, as a young foreign visitor, would make the most of this opportunity.

Gengozjan is also available to help us in any way with our practical needs; for example, the distances in Oak Ridge called for a car.

"I can help you buy one.", says Gengozjan. "The dealers here are all crooks and maybe I can protect you, if I know the guy".

Used cars can be found in open flat places around the valley. Gegozian takes me to a dealer, large and heavy, who practically lives in a car with open doors, which also functions as his office. The car I need must not be expensive, but must last at least two years. The dealer nods, but he is suspicious when he sees my companion, who goes around the cars on sale, hands tucked in his trousers' pockets, occasionally kicking the tires and shaking his head.

"Why did you take this man here?", mumbles the dealer. "I lost too many clients because of him!" I say:

"You must be mistaken! He's from the Biology Division!", but I appreciate Gegozjan's routine to keep the crooks on their toes. After a while, the dealer says:

"I have a *naice leedle Paaniac* here…(while pronouncing this sentence his half-done cigar travels from one to the opposite corner of his mouth, without being touched by his fingers)… Perfect for you!"

The little Pontiac, light blue, year 1952, sporting the first model of GM's hydraulic automatic shift and a rusty engine with eight cylinders and innumerable miles, is all but little, with heavy chrome sculptures on the front and three chrome stripes on the hood. As a first car, it is irresistible. I get it for $365.00. Two years later, I drive it to the dealer and he promises to sell it for me. But he is a crook, and this time I am unprotected. When a friend goes to see him to collect the money from the sale on my behalf, he finds the dealer lying in the light blue Pontiac, doors open. It is his new office!

E.H. Perkins is the oldest of the researchers and the kindest, always ready to go out of his way to help find a reagent or solve a problem. His face has a perpetual smile on it. However, about half way into my second year of stay, we notice that his smile has faded. At lunch, somebody asks:

"Is something wrong, E.H.?

"How did you know? I am a little overworked at home, with the children… Matilda has been sick"

"Is she better?"

"Not yet, but with God's help, I am sure she will recover soon".

A few days later, E.H. is silent. I ask him seriously:

"What happened? What is the doctor's diagnosis? I am an MD myself, I want to know!" Perkins seems relieved by my interest:

"She has a fever and shortage of breath. I did not call the doctor. You know, our Mormon religion teaches us to pray because God knows best, and he is the origin of salvation, and therefore of health, so we pray."

"Perkins!", I shout: "You must call the doctor immediately! I admire your faith, but God will not be offended if a doctor saves your wife using penicillin! If I were God I would actually be proud of you for putting my little fungus to good use!"

Makinodan, and the others, are worried by the unusual shouting they hear. I go on:

"Are the children not a sufficient argument? I have another: if God were really jealous of his miracles and offended, do you expect he would have permitted you to become a scientist and an immunologist, that is, a man free to think and reason"?

Perkins looks around. Everyone has the same look on their face. He will do the right thing. Even a bishop can act following his deep-down, rational conviction. Mrs. Perkins had a severe case of pneumonia. She was saved by penicillin. The Perkins' children became five, then six.

Bill Peterson. It came naturally, when speaking to this athletic young man, to talk sports. After the usual misunderstanding about football which, in the US is a completely different beast than soccer (and which, by the way, is MY football), Bill tells me he played basketball in school and volleyball in college. I play tennis. It turns out we only have ping-pong, or table tennis, in common. I tell him:

"It would be nice to play together some time! But where?"

Bill replies:

"It's the only thing missing in the division! I'll have to look in town". One month, two months, then three months go by. For a long while we are both too busy to think about it. But later on, a month before my departure, I say boldly:

"Bill Peterson, are you perchance afraid to be beaten? Ha! Ha! Ha!". He laughed with his all-white-teeth smile:

"I am not a sore loser, for sure! Dunnow about you Italians, but I was top three at Vanderbilt!"[130] Then, he becomes serious:

"I really did quite a search and found a number of places: in some, I can play; in others YOU can play. There isn't a single table where we can play together".

However, something is starting to move even in Tennessee in terms of civil liberties. Not with riding buses: one year ago the local bus service between the town and the hill on the outskirts, which The Manhattan Project had assigned to the black community has been discontinued. Recently, *The Daily Oak Ridger* has published a petition for equal rights, carrying the signatures of 95% of the researchers of the Biology Division (including all foreign visitors). A group of US scientists organizes a mixed, "black and white", dinner. When we, the visitors, offer to join, they politely refuse: "This is an American problem. We feel it is an American duty to manage it."

130. *Vanderbilt University* is in Nashville, Tennessee. It has 12,000 students, ranks high for academic rigor, has a most fervid campus, and excels in all sports.

Something Totally New

The day is April 12. We are now real Oak Ridgers, one part of the family will get the US Passport. Benedetto comes to our house near the forest, House F, Apt. # 4.

Benedetto is a regular guest of the Italian couples (a total of 4). Today there is no real meal: Anto feels tired, and I have driven to the Fried Chicken Shop, while Anto shows to Benedetto the results of last week's baby shower, when the invitation to the Welshons turned out to be a trick for me to take Anto to their house and deliver her to the Biology Division Wives: they opened the door shouting "Anto, dear!!!", took her from me and asked:

"Could you pick her up in two hours, please?"

Today, we begin eating the deep-fried chicken, which seems drier than usual, while Anto prefers to lie on the bed, leaving the door open to communicate. When the signal comes, my piece of chicken is not finished and for the rest of the evening I have the distinct feeling that something is stuck in my throat. Benedetto helps us down the stairs and into our 52 Pontiac, starts his Plymouth, and follows us. There is no traffic, and soon we reach the Methodist Church Hospital (the only hospital in Oak Ridge). A nurse rolls out a chair and Anto is taken in. The first room serves as waiting room. I thank Benedetto, who says:

"Let me know what happens! Call me!"

I enter the second door, rejoin Anto, and a few minutes later Dr. De Persio arrives. Now that we are all here we can proceed. However, the first lesson for new parents to learn is that normal childbirth is unpredictable in terms of waiting time, and the first time it takes really long. Luca is born at 4 AM when everybody, except him, is exhausted. About 4:30 I open the door to the waiting room, directed to the main exit. Somebody is sitting there.

"Benedetto! Still here! I can't believe it; six, seven hours!

We hug like bears: what manner of a friend! What an indelible memory!

The following morning, I rush to the hospital with a strange urgency to verify that everything I have experienced the night before has actually happened. I find Anto sitting in bed, with the baby who does not look too happy about what he sees and tastes. She has a tale:

"I will never forget his arrival from the nursery. This is the last room of the corridor from the nursery. At 7 a.m. sharp the cart of the babies comes into the corridor. It is on wheels, narrow, made of sections, each section contains a baby. Today they were eight, all eight crying desperately; after the first stop, the first baby exits from the choir. At each stop the crying party decreases by one. When only one voice remained, I knew it was Luca, and here he is!"

Franco is eagerly looking at the boy, but is interrupted by the arrival of the pediatrician; middle aged, salt and pepper hair, thick lenses, one eye seems out of axis.

"I am Dr. Frank Axelboum from Brooklyn. The one you chose for the first assessment of your little brat."

He picks up the baby, who tries to protest and cry, but is taken by surprise, like the parents: the doctor pinches the muscles, turns his head, shines a light into his eyes and ears. Suddenly he picks him up, two hands under the armpits, and sits him straight up on the empty bed. Then, he lets loose of him and steps back. The baby begins to lean to one side, the father makes the move forward in rescue but is kept back by the doctor and the baby falls softly and lies on the bed. This routine is repeated several times. The baby falls, but slowly and not always on the same side. Axelboum scribbles some notes in his book, and is ready to go. The new parents ask "So, Doctor, what are your conclusions?" Axelboum looks at the parents, then looks at the baby, fakes hesitation, and utters:

"I would keep him."

This, to us, is the New World.

Fig.92 - The baby cart, Methodist Hospital of Oak Ridge, Tennessee, 1960

Postface

The self-publishing option we use for *Memory Hound* introduces several novelties in book creation, distribution and in the relationship between authors and readers, increasing the choices of both, without causing drawbacks to neither.

Before, a book was a crystal, that would not change or grow until all copies of the first edition were sold to justify the second edition and so forth. Now the scarecrow of stored unsold volumes has disappeared thanks to faster printers, able to create any number of copies in a day, and enabling *printing on demand*, where only books already bought are printed. Books are becoming something live and dynamic, allowing updates and changes to the text, especially when the field is Science or related to Science, like in the present case.

Suppose a reader has new information, and wants, to discuss it with the author, or suppose that an article appears in *Nature*, that would render obsolete a description in this book: in both cases the recognition of the reader's opinion, and the correction on our page will be rapid, and the updated version will be printed from that point on, with great rise of value.

Another example: Do you remember that, in the Chapter 10 (*Stefania*), I describe the *self-recognition* of one's own memories (varying with the age of the *Rememberer* at the fixation moment), and lamenting the lack of similar observations of other people's experiences? Well, with a book able to grow by a series of new self-recognition data, this is a real collaboration, and with enthusiasm I will ask you to participate. It is in the character of a *live* book to create a space for materials related to chapters for your browsing, documents and drawings, alternative or additive to those included in the text.

FC

Acknowledgements

To proceed from a manuscript, self-typed or handwritten, you need all kinds of help: the first is encouragement, the rest spans from professional communication to computer savvy. I found all this and more in Valentina, my youngest daughter and "kindred spirit". Born in Stockholm, graduated in Letters and Sculpture in Rome, practicing Applied Art at the Opera, she is a fixer with a smile. I gratefully acknowledge her rational editing and genial art direction of the book.

Fig.93 – Franco Celada and daughter Valentina

Five years ago, Paolo Mignatti, an Italian cell biologist and basic oncologist, was assigned to the sixteenth floor of 311 East 17th street, the Hospital for Joint Diseases in New York City. I had been there since 1983. Was it simply good chance or was it scripted by Frank Capra? Despite a twenty-year age difference, we began chatting, discussing life and lunching together at the cafeteria, often celebrating five p.m. with a mug of hot chocolate from Turin, as if time did not exist: becoming friends. Paolo was curious about my chapters, and from that point on became the first reader and the last reviewer of all my English pages. I gratefully acknowledge his encouragement and critical judgment, within a beautiful friendship.

Fig.94 - Franco Celada with friend and colleague Paolo Mignatti at the Cafeteria of NYU - Hospital for Joint Diseases

I am indebted to all, Relatives, Friends, Colleagues and Pupils, who contributed to "hound", enriching the book with their version of collective memories with sufficient emotion to be remembered, at times suggesting new facts or interpretations altogether. The travels and meetings with many of you have given me one pleasure that is unique to old age and that I cherish.

Franco Celada

Figure Index

Guido's Legacy

Both Celada and Longoni, with the families growing, moved away from the apartment house of Via Eustachi to larger quarters in the neighborhood. The Longoni did it in 1934. The new address, Via Stradella 1, was so near, that no movers were needed. A dinner and dancing party was organized. Many friends came, bringing flowers and presents. Guido, being an excellent oil painter and genial with caricatures, brought a 2 x 6 ft drawing, illustrating the Longoni family moving. As soon as he opened it and pinned it to the wall, there was a roaring laughter. Stefania, who was six at the time, is a witness (Franco was four and was not invited) and she knew the drawing had been carefully stored for years by her parents. Only now, in 2019, she found them in the attic. The medium is colored pencils and crayons on cardboard paper. The freshness of the original drawing, after 85 years, is astonishing.

Guido Celada's drawing on the previous page is a testimony to the Celada-Longoni lifelong friendship and to publish it is a real joy. However, the black and white printing doesn't do justice to the artwork. For this reason, if any reader should be interested, Memory Hound will send a PDF color copy of the image. Requests can be sent to: memoryhound@gmail.com.

About the Author

Franco Celada was born in 1931 in Milan, Italy. After Classical Studies at the *Liceo Gonzaga*, he earned his MD degree and specialized in Internal Medicine at the *University of Milan* and started a two-year postdoctoral engagement at the *Oak Ridge National Laboratory* in Tennessee to study radiation immunology under the guidance of Takashi Makinodan. In 1962, he left medical practice and embraced basic research accepting the position of Senior Assistant in the *Department of Human Genetics* of the *University of Torino*, focusing on *Transplantation Immunology* with Chairman Professor Ruggero Ceppellini. Celada left after two years, invited to work at the *Karolinska Instititute* in Stockholm by Professor George Klein, an authority in Tumor Immunology. The seven years in Stockholm proved important, initiating the studies on *Immunological Memory*. At Karolinska, he discussed the thesis for his Ph.D. in Immunology *cum laude*, based on his work in the US and Sweden, earning a *Docent* position, which eventually lead him to become Professor in 1969.

After working with Jacques Monod in Paris at the *Institute Pasteur* with a fellowship of the French Government, he became Director of Research at *The Italian Research Council* (C.N.R.) and Head of the Immunology Section of The *Laboratory of Cell Biology* in Rome, Italy in 1971. After four intense years at the lab, a sabbatical was due. Two scientists of UCLA, Eli Sercarz and Irving Zabin, invited Celada to work with them for a year. In 1976, he was nominated Professor of Immunology, at the *University of Genoa*, Italy, where he was also in charge of the histocompatibility typing for the *Kidney Transplantation Service* at the *San Martino* hospital. From 1982 to 1987, he served as the Coordinator of the *Italian Ph.D. Program in*

Immunological Sciences of the *University of Genoa*. In 1987, after a sabbatical at *Mount Sinai Hospital*, he became Senior Immunologist at the Department of Rheumatology of the *Hospital for Joint Diseases* in New York City. From 1985, Franco Celada divided his time between the US and Italy until 2006, when he was nominated Emeritus by the *University of Genoa*, and moved to a fulltime activity at the *NYU Medical School* as Research Professor of Medicine, where he is still active.

Throughout his career, Franco Celada has been involved in international educational programs. He was a founder and first President of the *Italian Cooperation Society in Immunology* in 1973 and was nominated member of *EMBO* (European Molecular Biology Organization) and member of the *EMBO Course Committee* in1974 and part of the Editorial Board of EMBO Journal since 1991. He is member of the *American Association of Immunologists* since 1989 and Chairman of *Education Committee of IUIS* (International Union of Immunological Societies) from 1986 to 1994. From 1987, he founded a European Committee Group for the creation of a *European Doctorate in Bio-Technology (*EduBT) involving twenty Universities and serving as secretary of EduBT from 1987 to 1993. For six years he was Chairman of a European Group promoting and researching *BioEthics*.

During the five decades of scientific research across the world, Franco Celada's research focused on *Autoimmunity* (finding Anti-DNA antibodies in Lupus), *Histocompatibility* (finding radiation-resistant cellular responses in mice), *Immunological Memory* (measuring its decay in adoptively transferred cells) and *Antibody-Mediated Conformational changes of proteins*, which control enzymatic function or, e.g., the membrane penetration site for viruses.

Celada's philosophy of research pivots around the idea of cooperation and exchange among disciplines, which has helped reveal the Immune System's nature of *Cognitive System*, acting like a second, un-conscious mind of the body. The new conception of the Immune System makes further collaborations across disciplinary barriers easier and useful.

In the late 80's, Celada met IBM Astrophysicist Philip Seiden, with whom he developed *IMMSIM* (*http://immsimteam.med.nyu.edu*) a discrete computational model of the entire Immune System, an Automaton on which virtual experiments are run to generate predictions, potentially becoming scientific facts, when validated in vivo. Their first publication came out on *Immunology Today* in 1992 and the Celada-Seiden model grew in size, scope and results over the years supported by a long list of presentations and articles. After the sad and unexpected loss of Seiden in 2000, the modeler's role was picked up by Physicist Roberto Puzone and a fruitful collaboration with Liisa Selin of Umass was aimed at the central problem of *Antiviral Memory*: the escape of the invaders by mutation. Its novel, "investigative" design was appreciated by immunologists.

Today, 29 years after its inauguration, IMMSIM is still used. The most recent article was published in July 2019 on *Frontiers of Immunology* by Filippo Castiglione, Dario Ghersi and Franco Celada: a case-study of the potentially dangerous reaction of the Memory anti-Naïve cells to a mutated virus (the MaN phenomenon), in which IMMSIM's predictions reveal the presence of self-regulatory mechanisms in the Immune System (in this case *Attrition*) in order mitigate the response, further clarifying the so called *Antigenic Sin*, that has been puzzling Virologists and Vaccinologists for more than half century.

PUBLISHED BOOKS

1. Franco Celada (Editor). *Metodi dell'Immunochimica.* Boringhieri, Milano, 1975.
2. Franco Celada, Verne N. Schumaker and Eli E. Sercarz (Editors). *Protein Conformation as an Immunological Signal.* Plenum Press, New York and London, 1983.
3. Franco Celada. *Le Difese Immunitarie: come l'organismo riconosce se stesso e il mondo esterno.* Le Scienze SpA. Editore, Milano, 1986.
4. E.E. Sercarz, F. Celada, N.A. Mitchison, T.Tada (Editors). *The Semiotics of Cellular Communication in the Immune System.* Springer Verlag, Berlin and Heidelberg, 1988.
5. Franco Celada (Guest Editor). *Non-Conventional T Cytotoxicities*, Vol.4, No.2, Intl. Reviews of Immunology. Harwood Academic Publishers, 1989.
6. Franco Celada (Editor). *Il Sistema Immunitario* (Le Scienze, Quaderno 55). Le Scienze S.p.A., Milano, 1990.
7. Franco Celada and Benvenuto Pernis (Editors). *T-Lymphocytes: Structure, Functions, Choices.* NATO A.S.I. Series, Vol. A233. Plenum Press, New York and London, 1992.
8. Franco Celada (Editor). *La Nuova Immunologia.* Le Scienze S.p.A., Milano, 1992.
9. Ulrich Behn, Franco Celada, Philip Seiden. Modelli *Matematici in Immunologia.* Frontiere della Vita, pag. 573-589. Enciclopedia Italiana Treccani Editore, Roma, 1988.
10. Filippo Castiglione, Franco Celada. *Immune System Modeling and Simulation.* CRC Press, Boca Raton, London, New York, 2015.

SELECTED PUBLICATIONS

(Selected from a total of 120 peer-reviewed publications)

1. Ceppellini R, Polli E, Celada F. *A DNA-reacting factor in serum of a patient with lupus erythematosus diffusus*. Proc Soc Exp Biol Med.1957; 96:572-4.
2. Celada F, Makinodan T. *A new model to study hematopoietic transplantation antigens*. J Immunol. 1961; 86:638-45.
3. Celada F. *Quantitative studies of the adoptive immunological memory in mice*. I. *An age-dependent barrier to syngeneic transplantation*. J Exp Med. 1966; 124:1-14.
4. Celada F. *The cellular basis of immunological memory*. Progr Allergy 1971; 15:223-67.
5. Rotman MB, Celada F. *Antibody-mediated activation of a defective beta-D-galactosidase extracted from an Escherichia coli mutant*. Proc Nat Acad Sci USA 1968;60:660-7.
6. Celada F, Ellis G, Bodlund K, Rotman B: *Antibody-mediated activation of a defective β-D-galactosidase*. II. *Immunological relationship between the normal and the defective enzyme*. J Exp Med. 1971; 134:751-64.
7. Celada F, Cambiaggi C, Maccari J, Burastero S, Gregory T, Patzer E, Porter J, McDanal C, Matthews T. *Antibody raised against sCD4-rgp120 complex recognizes the CD4 moiety and blocks membrane fusion without inhibiting CD4-gp120 binding*. J Exp Med. 1990; 172:1143-50.
8. Celada F, Seiden PE. *A computer model of cellular interactions in the immune system*. Immunology Today 1992; 13:56-62

9. Kohler B, Puzone R, Seiden P, Celada F. *A systematic approach to vaccine complexity using an automaton model of the cellular and humoral immune system.* Vaccine 2000; 19:862-76.
10. Cheng Y, Ghersi D, Calcagno C, Selin LK, Puzone R, Celada F. *A discrete computer model of the immune system reveals competitive interactions between the humoral and cellular branch and between cross-reacting memory and naïve responses.* Vaccine. 2009; 27:833-45. PMCID: PMC3905836 http://www.sciencedirect.com/science/article/pii/S0264410X08016484
11. Castiglione F., Ghersi D., Celada F., (2019) *Computer Modeling of Clonal Dominance: Memory-Anti-Naïve and its Curbing by Attrition.* Frontiers in Immunology. *10:1513. doi: 10.3389/fimmu.2019.01513*

Index

Made in United States
Cleveland, OH
05 November 2026

47669892R00220